INDIAN GEOMORPHOLOGY

BY THE SAME AUTHOR

Perspectives in Geomorphology (4 Vols.) Rs. 800 per set

Vol. I Recent Trends

Vol. II Quantitative Fluvial Geomorphology

Vol. III Applied Geomorphology

Vol. IV Essays on Indian Geomorphology

The Physiography of the Lower Chambal Valley and its Agricultural Development, Rs. 60

Ravine Erosion in India, Rs. 50

Tropical Geomorphology: A Morphogenetic Study of Rajasthan Rs. 300

FORTHCOMING

Human Impact on Environment (Concept's International Series in Geography No. 6)

INDIAN GEOMORPHOLOGY

(Selected papers from the 2nd Conference of the Indian Institute of Geomorphologists on Environmental Geomorphology)

Edited by
H.S. SHARMA

Foreword by
S.N. SINHA

CONCEPT PUBLISHING COMPANY, NEW DELHI 110059

ISBN 81-7022-344-X

First Published 1991

Published and Printed by
Ashok Kumar Mittal
Concept Publishing Company
A/15-16, Commercial Block, Mohan Garden
NEW DELHI 110059 (INDIA)

Lasertypeset by
Microtech Advance Printing Systems (Pvt.) Ltd.
H-13, Bali Nagar
NEW DELHI 110015

To
My wife
Maya
for her
sacrifices
and
encouragement

Foreword

IT GIVES me great pleasure in introducing the book entitled *'Indian Geomorphology: Environmental Geomorphology' - I*. The book contains selected research papers of the Second Conference of the Indian Institute of Geomorphologists, held at Jaipur from 2-4 November, 1989. It was the proud privilege of the University of Rajasthan to host this conference on the theme Environmental Geomorphology. The editor of the book, Dr. H.S. Sharma has meticulously collected the material and has rationally organised and edited. Dr. Sharma himself has made seminal contribution in the field of geomorphology.

For a long time, geomorphology was considered a subject comprising mainly academic studies of the types and origin of landforms. Only relatively recently it has found many applications, particularly in various fields of environmental resource surveys. As a result, geomorphology has received much attention from growing numbers of scientists and planners. Swelling in numbers, there are non-geomorphologists, including soil scientists, engineers, geologists, regional and urban planners, etc. The development of the subject has, therefore, gained new momentum with respect to its scientific foundations and its methods of research.

The diverse field of science covered by geomorphology encompasses many subjects, particularly since the recent development of the past few decades. Basically, four major aspects of geomorphology can be distinguished as (i) static geomorphology – concerned with actual landforms, (ii) dynamic geomorphology – concerned with processes and the short-term changes so caused landforms, (iii) genetic geomorphology – concerned with long-term development of relief, and (iv) environmental geomorphology – concerned with the landscape ecological links between geomorphology and order disciplines of the land.

India has many problems which are structural, hydrological and agronomical in nature. Among these hydrological, agricultural, drought and desertification are most fascinating and confronting. The arid and semi-arid regions are awaiting for the permanent solution of water scarcity. On the other hand, the humid parts of the country are suffering from problem of floods which also need consideration. In the irrigated tracts of the Punjab plain, the Ganga-Yamuna Doab, and the Lower Chambal Valley due to heavy nature of clay soil the problem of water-logging has emerged. The designing and construction of Indira Gandhi Canal in a fragile arid ecosystem of Rajasthan without proper understanding the dynamics of geomorphological form and processes has resulted in development of large-scale salinisation in Suratgarh and Hanumangarh areas of Rajasthan. Reclamation of saline and alkaline soil is another threatening problem that has emerged with water-logging in various irrigated tracts of the country. The constant increase of ravine lands every year poses a serious problem not only in the landuse planning but in the irrigation planning as well. In order to solve these problems, geomorphological studies and mapping of various regions and sub-regions of India are very essential. Geomorphological knowledge, methods, techniques and way of approach, can be used in combating all these problems which may further help in land resource development and water resource management.

As a reference I would like to refer the Report *The Limits of Growth of the "Club of Rome" that opened the eyes of many people*, both scientists and layman. More specific in several respects was the report of the *World Commission on Environment and Development*, the *Brundtland Report* published in 1987. The keywords here is "sustainability". The call for sustainable development and the remedial measures are required. What is needed, in fact, is a fundamental reorientation of our priorities within the structure of social and economic goals, with the concept of economic growth being replaced by environmental management. During an International Workshop on Long-term World Development held in Sweden in 1986, several scenarios were sketched for the period 1975-2075, each related to a specific mode of human response. The workshop's report was entitled 'surprising future' (1987) and it makes surprising reading indeed. We have suddenly come to realise that it is extremely difficult – with our growing world population and our growing economic ambitions – to live in harmony with our environment. What has been called "the rape of the earth" is taking on frightening dimensions. The human race, through its non-sustainable activities, has become a major causative factor of global change which, unfortunately usually takes a negative degradational form. Global change has many aspects, including the erosion of biological diversity, acidification of soils, pollution of groundwater by nitrogen reaching in agricultural areas etc.

Since all these issues, as I pointed out above are of multi-disciplinary nature, it is essential for geomorphologists at this stage to participate in multi-disciplinary organisations such as the International Organization for Quaternary Research (INQUA) and the International Geomorphical Union (IGU) if they are to make a successful contribution to the complex research in the area of environmental hazards mitigation. All in all, there can be no doubt that geomorphologists can, in co-operation with colleagues from other disciplines, contribute to an appreciation of environmental hazards, particularly when applying remote sensing technology, by indicating endangered areas through natural hazards zoning and monitoring process involved.

Thus, the book containing 29 research papers on various aspects of environmental geomorphology is indeed a service of geography to thinkers and to planners. I have great pleasure in commending the book to all those who are interested in the study of environmental problems and of several aspects of resource management. Dr. Sharma deserves our appreciation for his dedicated efforts in bringing out this volume in the published form.

University of Rajasthan
Jaipur

S.N. SINHA

Preface

THE ENVIRONMENTAL geomorphology concerns with the landscape ecological links between geomorphology and neighbouring disciplines or elements of the land. Geomorphologists have studied environmental hazards and problems since the early parts of the present century, but until late 1950s their research focused almost exclusively on theoretical aspect. During the past 30-35 years, however, there has been a growing interest in applied geomorphology, and today research on morphodynamic processes that triggered of by man's interference dominates this branch of geomorphology. One manifestation of this is the organisation of two International conferences on geomorphology encompassing the theme of ecology and environmental management held at Manchester, U.K. in September 1985 and Frankfurt in September 1989. The International Symposium on Aerospace Survey and Natural Hazards Predication and Reduction at ITC, Enscede, Netherlands in September 1989 has witnessed the growing importance of this branch of geomorphology in combating present environmental problems. In recent years new concepts of geomorphology have mushroomed leading to a completely different appreciation to the environment. This has also a profound effect on the application of geomorphology which now are increasingly geared to matters of global change and environmental hazards reduction and prediction.

Environmental changes and processes are studied in a variety of disciplines other than geomorphology. Moreover, studies within geomorphology owe a considerable depth to and are often closely linked with these other disciplines, which include geology, pedology, geography, civil engineering, hydrology, forestry etc. Thus the study of environmental problems and management is truly an interdisciplinary nature. At the heart of environmental science is geomorphology, in so far as it is concerned with morphological forms and processes, it alone seeks to integrate the more specialised knowledge generated by the other disciplines.

Because the scientists conducting research on environment, work either within or on the margins of several disciplines, they bring to the subject different sets of skills, experiences, and approaches. For this very reason it is highly desirable to interact and exchange views and informations. However, precisely because they are associated with desperate disciplines, they have few opportunities to meet and discuss their research findings. The second annual geomorphology conference was therefore concerned with the objective of bringing together representatives of the various disciplines. The goals of the conference were to provide a forum for the presentation of current research on environmental geomorphology, to facilitate the exchange of ideas and the dissemination of new developments across disciplinary boundaries and to generate new initiatives for further research and co-operation.

During the International Symposium on Geomorphology and Environmental Management held from 17th to 20th January, 1987 at the University of Allahabad, the Indian Institute of Geomorphologists (IIG) was formed. The first conference of the Indian Institute of Geomorphologists was held from 28th to 30th October, 1988 at the Geography Department of Andhra University, Visakhapatnam under the dynamic convenership of Prof. R. Vaidyanadan, a noted geomorphologist of this country. The second conference of IIG was hosted by the Department of Geography, University of Rajasthan, Jaipur from 2 to 4 November, 1989. It was attended by about 100 delegates from all over the country.

The present volume contains proceedings of the second conference and national symposium on environmental geomorphology. The book contains twenty-nine papers on various aspects of applied and environmental geomorphology.

The first chapter on the status of applied geomorphology is by K.R. Dikshit. He demonstrates that the significant features of applied geomorphology is its late arrival, attributed to excessive obsession of geomorphologists with classical geomorphology and denudation chronology for a long time. The duality of a structure based and a climate-based geomorphology ignored the application of their findings to different areas of economic activity. He further stressed that both the basic and applied nature of the subject are important and complementary, the welfare of the humankind is the focal point in the development of all disciplines and geomorphology is no exception. He questions should we not then, practice the art of geomorphology to redeem our existence? M.L. Dewan in his key-note address has discussed the application of environmental geomorphology and soil sciences to ravine reclamation -- a national problem of a great magnitude. He stressed for the detailed geomorphological studies of ravine areas for evolving a strategy for reclamation of degraded areas. Involvement of people in reclamation of ravines will also go a long way to reduce this menace. P.C. Vats and Surendra Singh have studied the ecosystems and environmental hazards of an arid environment with special reference to Barmer district. Geomorphological aspects of the processes of desertification as a case study of Sambhar lake and its environs have been discussed by H.S. Sharma. He has elucidated a number of morphodynamic processes enhanced and sometimes triggered off by anthropogenic factors.

R.K. Rai and R.C. Joshi in their paper have discussed land capability and land management with a ecological viewpoints of the Central Himalaya. Land evaluation and associated landuse systems in semi-arid Morel Basin has been studied by R.D. Doi, R.M. Pofali, J.L. Sehgal and M. Swaminathan in their paper on problems and potentials of pedo-geomorphic surfaces in the Purna Basin have emphasised that scientific management of land, soil and water is imperative for optimum utilisation of these resources to the best of their capabilities without disturbing the ecological balance. The problem of flood and bank side erosion are typical chronic problems in the flood plains of the lower Damodar Basin in West Bengal. Probhat Kumar Sen has studied this aspect in great detail. He pointed out that in spite of the gainful consequences of soil conservation, flood mitigation, irrigation in the upper and lower catchments, the flood problems of the basin have not been completely ameliorated.

The introduction of canal irrigation in the fragile arid ecosystems without taking into cognizance the morphological character of the land and the progressive mismanagement of land-water resources, can hasten the process of desertification including, waterlogging, salinisation and other hazards. S.C. Dhiman and H.S. Sharma studied the waterlogging problem of Shri Ganganagar district with the help of remote sensing technique and traced the temporal changes in land conditions. S.M. Bhamare in his paper emphasised to bring out the relation between groundwater and calcretes in the Panzara Basin. V.K. Asthana and Karan Vir in their paper have identified morpho units of the Pervikhala Drainage basin applying cluster analysis. They stressed that the degree of generalisation in cluster analysis has led to rather large simplification of information, and overcasts the details of topographic fluctuations. Patil, Y.V. in his paper studied occurrence of sandstone in the Tapi flood plain. Rock withering and soils of the Mawsynram area of east Khasi hills, district Meghalaya has been studied by Girish Chandra Panda and R.K. Rai.

Srinivas Madabhushi in his article has studied the drainage, lithology and structure of the north western part of Rajmahal hills, Bihar. Ravindra K. Pande in his paper has discussed slope instability problems in Karmi area of central Himalaya. Gopal Lal Gupta has studied the role of slope in landuse of a Bamanwas tahsil. Drainage density in relation to rainfall has been analysed by Yaseen Khan. Devendra Kumar in his paper has discussed development of hill slopes of a granite hill.

A.B. Mukerji in his detailed study on the Punjab Piedmont Pediments attributed their genesis to faulting and sheet erosion. According to him presently, the pediments appear, more or less, dead features, fossil in character. They are features of both climatic geomorphology and structural geomorphology. Another paper on applied geomorphology of Ladakh and Trans-Himalayan region by R.C. Pathak discussed glacio-fluvial features of the region and their applied significance. Manjari Bhattacharji and V.A.V. Raman in their paper on geomorphic observations in the middle reach of the Parvati river of Kulu district, Himachal Pradesh have pointed out that the geomorphic evidence are indicated of neo-tectonism within this region.

Remote sensing technique has been recognised as a significant and affective tool for identification and mapping of geo-environmental features. In the case study of Tirupati area P.S. Dhinva, Jebakumer, K.L. Majumdar have emphasised the suitability of image processing techniques for identification of geomorphological features. Surendra Singh, Amalkar and P.C. Vats in their paper demonstrate role of remote sensing in environmental management in case of Rajasthan. Alexander S. Shestakov in his study has analysed desertification in Rajasthan specially the central Aravallis with the help of satellite imagery. M.S. Nathawat, M.P. Punia, R.R. Saini, Sanjay Raj, Dharmendra Sharma, Anirudh Sahni and A.K. Sinha have emphasised in their papers the role of landsat imageries in environmental conservation.

K.N. Prudhvi Raju and Ahmed Raza Khan in the paper on a conspicuous low stand of sea level in the inner shelf zone off Visakhapatnam coast, Andhra Pradesh have discussed various features which indicate a conspicuous low stand of sea level at 10-11 metres depth. Abha Mathur has studied longitudinal profiles of the selected river basins of Rajasthan in her paper. The last paper of this volume by S.C. Kalwar and P.L. Vijai attempts to study the relationship between morphometric parameters and net sown area of the Khari basin, Rajasthan.

During the past four decades, the advances made in environmental geomorphology have furthered our understanding of channel properties, its geometry and hydrology, the discharged characteristics of the rivers, the sediment load, the slopes and slope failures, floods, the landslides, the impact of waterlogging, resulting from impeded drainage or excess of irrigation, and the host of other phenomena. Although environmental geomorphology has come a long way in four decades, it is still far from maturity, and much work remains to be accomplished. Many challenges lie ahead, and it is hoped that the present volume will be of help in meeting such challenges.

University of Rajasthan
Jaipur 302004

H.S. SHARMA

Acknowledgements

THE HIGH quality of the chapters in this volume is due to at least in part to the diligent efforts of the following individuals who gave freely of their time to review the manuscripts: Prof. N.L. Gupta, Prof. A.B. Mukerji, Prof. K.S. Sivasamy, Prof. Indira Pal, Dr. S.C. Kalwar, Dr. M.L. Sharma, R.D. Gurjar. To these people I express my sincere appreciation for their help and generosity.

The second annual conference of the Indian Institute of Geomorphologists, the proceedings of which are contained in this volume, was supported by grants from the University of Rajasthan, Jaipur out of unassigned grant of the University Grants Commission. The grant permitted participants to travel from different places to attend the conference and symposium and contributed to its ultimate success. For this I express my heartiest thanks and appreciation to Prof. S.N. Sinha, Vice-Chancellor, University of Rajasthan, for his blessings and generosity. Thanks are also due to my colleagues of department of geography, for their whole hearted co-operation in successful realisation of this academic event in November 1989.

H.S. SHARMA

Contents

List of Contributors

K.R. Dikshit
Department of Geography, University of Poona, Pune (Maharashtra)

M.L. Dewan
Himalayan Surakshan Ashram, Rani Chauri 249199, Tehri Garwal Distt. (U.P.)

P.C. Vats
Basic Resources Division, Central Arid Zone Research Institute, Jodhpur

Surendra Singh
Basic Resources Division, Central Arid Zone Research Institute, Jodhpur.

Amal Kar
Basic Resources Division, Central Arid Zone Research Institute, Jodhpur

H.S. Sharma
Department of Geography, University of Rajasthan, Jaipur 302 004, (Rajasthan)

R.K Rai
Department of Geography, North Eastern Hill University, Shillong, Meghalaya

R.C. Joshi
Department of Geography, North Eastern Hill University, Shillong, Meghalaya

R.D. Doi
Department of Geography, University of Rajasthan, Jaipur (Rajasthan)

R.M. Pofali
National Bureau of Soil Survey and Landuse Planning, Amravati Road, Nagpur - 440 010.

J.L. Sehgal
National Bureau of Soil Survey and Landuse Planning, Amravati Road, Nagpur - 440 010.

M. Swaminathan
National Bureau of Soil Survey and Landuse Planning, Amravati Road, Nagpur - 440 010.

Probhat Kumar Sen
Department of Geography, Burdwan University, Burdwan 713 104, West Bengal

S.C. Dhiman
Central Ground Water Board, Bani Park, Jaipur

S.M. Bhamare
Department of Geography, S.S.V.P.S. Science College, Dhule - 424005

V.K. Asthana
Department of Geography, Kurukshetra University, Kurukshetra

Karan Vir
Department of Geography, Kurukshetra University, Kurukshetra

Y.P. Patil
Department of Geography, Jai Hind College, Dhule

Girish Chandra Panda
Department of Geography, Union Christian College, Barapani, East Kashi Hills District, Meghalya

Srinivas Madabhushi
Geological Survey of India, Visakhapatnam

Ravindra K. Pande
Department of Geology, Kumaun University Campus, Almora - 263 601 (U.P.)

Gopal Lal Gupta
Research Scholar Department of Geography, University of Rajasthan, Jaipur 302 004

Yasin Khan
Research Scholar, Centre for Regional Development, Jawahar Lal Nehru University, New Delhi

Devendra Kumar
Department of Geography, R.D. College, Sheikhpura (Munger)

A.B. Mukerji
Department of Geography, Panjab University, Chandigarh, Punjab.

Lt. Col. R.C. Pathak
Research and Development Eastt. (Eg Engrs) Pune - 15

Manjari Bhattacharji
Research Scholar, Department of Geography, University College of Science - 35, Ballygunge Circular Road, Calcutta 700 019

V.A.V. Raman,
Research Scholar Department of Geography, Delhi School of Economics Campus, University of Delhi, Delhi 110 007

P.S. Dhinwa
IPDPG, Space Application Centre, Ahmedabad - 340 053

Jebakumar
IPDPG, Space Application Centre, Ahmedabad - 340 053

K.L. Majumdar
IPDPG, Space Application Centre, Ahmedabad - 340 053

Alexander S. Shestakov
Department of Geography, State Moscow University, Moscow (USSR)

M.S. Nathawat
BM Birla Science & Technology Centre, Jaipur

M.P. Punia
BM Birla Science & Technology Centre, Jaipur

R.R. Saini
BM Birla Science & Technology Centre, Jaipur

Sanjay Raj
BM Birla Science & Technology Centre, Jaipur

Dharmendra Kumar
BM Birla Science & Technology Centre, Jaipur

Anirudh Sahni
BM Birla Science & Technology Centre, Jaipur

A.K. Sinha
Department Geology, University of Rajasthan, Jaipur - 302 004

K.N. Prudhvi Raju
Department of Geography, Jamia Millia Islamia, New Delhi - 110025

Ahmad Raza Khan
Department of Geography, Jamia Millia Islamia, New Delhi - 110025

Abha Mathur
Amita Colony, Jaipur.

S.C. Kalwar
Department of Geography, University of Rajasthan, Jaipur, (Rajasthan).

Prabhu Lal Vijay
Department of Geography, Government College, Tonk, (Rajasthan)

1

The Status of Applied Geomorphology*

K.R. DIKSHIT

THOUGH THERE is a considerable corpus of scientific literature, on both, theoretical as well as applied geomorphology, one must admit that the experience and the knowledge gained through decades of research, field-observations and the laboratory experiments have not found their appropriate application in the areas of resource and environmental management, prevention of natural catastrophes and the monitoring of man's economic activities, particularly those involving land and water as their vital components. The reason lies in the history of the discipline, since impetus for its growth came necessarily from the inquisitiveness of the scholars trained in scientific methods, where deductive reasoning was the basic method to understand the landscape. This was a compelling necessity as the functioning of most of the geomorphic processes, with the exception of those having an episodic character, was extremely slow and required a very long-term observation.

The replication of geomorphic processes in the laboratory always poses serious problems, not the least of which is the problem of scale. A change of scale or miniaturisation could not be accepted since it meant a qualitative distortion of the phenomena and militated against the concept of threshold. Besides, the number of variables influencing a geomorphic phenomenon is not only large but almost limitless, often impossible to reproduce in the laboratory. The episodic and unpredictable, events make the picture still more tangled. Notwithstanding these difficulties, as more and more information

---

* Presidential address delivered on 2 November, 1989 at the Second Conference of the Indian Institute of Geomorphologists held of Jaipur from 2 to 4 Nov., 1989.

accumulated, it became clear that the landforms fall into a pattern and reflect the impact of contemporary climate and the lithological or structural milieu of their origin.

The analytical work on specific areas, type landforms, or in a specific climate, produced a better correlation of geomorphic events, resulting from structural changes or climatic characteristics, giving rise to a system of landforms in our perception, like the landforms of cold environment, as against those in the Tropies, or the landforms in the cratonic areas juxtaposed to those in recent sediments. Identification, genesis and dating remained the principal preoccupation of the geomorphologists for a long time. Any study that ignored landforms was treated with scorn. It is only after the Second World War that the focus shifted to the study of processes. The latter necessitated a peep into the areas, peripheral to the processes, lying in other cognate disciplines. It also brought home the reality that the geomorphic processes and, in some measure, even the landforms have an inescapable impact on human activities which, in turn, operate as a geomorphic factor. In fact, the anthropocentric factors are so effective that they function as a triggering device and bring in their wake a chain of cascading sequence in the geomorphic processes. This interrelation between the geomorphic processes and the human-economic activities is what forms the basis of applied geomorphology.

The understanding of this relationship has deepened, and as a consequence applied geomorphology has won recognition among many specialised branches of the discipline that have emerged over the years. Though there has always been a mention of the application of geomorphic principles to engineering and construction projects, ground water, location of dam-sites, flood control, management of desert fringe, land slides and soil erosion, these themes were together discussed for the first time by Thornbury (1954). Since then a large number of research papers and books have appeared that focus attention on applied geomorphology.

Late Arrival of Applied Geomorphology

The significant feature of Applied Geomorphology is its late arrival, attributed to excessive obsession of geomorphologists with classical geomorphology and denundation chronology for a long time. The duality of a structure-based and a climate-based geomorphology, each emphasising its own importance, promoted, in depth, the study of zonal landforms ignoring the application of their findings to different areas of economic activity. And wherever some degree of geomorphological knowledge was applied, it was invariably related to topographic conditions. The selection of dam sites, the laying of the canal and roads and the detection of groundwater were some of the earliest avenues in which the help of a geomorphologist was sought. Placer deposits associated with river channels often required a search for palaeochannels that could make the exploration of minerals occurring in placers easy. Gold, platinum, tin and diamond are some of the minerals usually obtained from alluvial placers. The detection of buried placers and, in turn, buried channels, brought geomorphology into picture. Decades earlier, a civil engineer or in some cases a geologist substituted a geomorphologist, as they do today, since geomorphology is considered a branch of geology as much as a branch of geography, and it is only lately that the discipline has started asserting its importance within the framework of earth sciences.

The Widening Scope of Applied Geomorphology

During the past four decades, the advances made in geomorphology have furthered our understanding of channel properties, its geometry and hydrology, the discharge characteristics of the rivers, the sediment-load, the weathering profiles and the sequences of their horizons, the slopes and the slope failures, the landslides, the impact of waterlogging, resulting from impeded drainage or the excess of irrigation, and a host of other phenomena. The deserts are much better understood today and so are the coastal areas. The geomorphologists have, and can acquire, an idea of the impact these processes will have on the life and economy of the concerned regions. A very conscientious, and a highly specialised geomorphologist may dismiss the impact aspect of the geomorphic phenomena, events and processes, as something beyond the scope of his work, yet one may justifiably argue that to ignore the impact aspect or the application of the understanding of the geomorphic processes is a logical absurdity. It does not appear rational to understand the causes of floods and then stop there without trying to work out the strategies to prevent them.

Scope of Applied Geomorphology

Hart (1986) divides Applied Geomorphology into three branches. The first concerned with natural hazards, the second with environmental management and the third with the evaluation of the resources. This three-fold classification is too simplistic and focuses heavily, almost exclusively, on the environmental hazards and management. The concern of applied geomorphology could be much more positive than the crisis management of the environmental resources. It has a wider role in so far as it is a question of harnessing and utilising the land and the water resources of a region. In very broad terms, geomorphology is concerned to redeem itself by being in the service of the society as expressed by Brundsen (1955).

"The present political and socio-economic world context for geomorphology is the disparity between the wealth of nations, a rapid increase in the number of people living at subsistence levels, and a determination that the standard of living of the people of the developing countries should be raised." This concern is supported by the acquired expertise of the geomorphologists in understanding the behaviour of the earth's material, processes and their capability in participating and assisting in the discovery, assessment and wise management of the earth's finite resources, to prevent environmental deterioration and natural hazards. The appreciation of the role of geomorphology, during the last couple of decades has resulted more from the environmental concerns of the affluent societies, then from the understanding of the perceived natural hazards which always occurred as episodic events in the history of the earth.

Areas of Application

One of the early direct application of geomorphology involved Terrain Classification in Australia. This was as much an advance in geomorphic mapping with facets of 'land-systems' and land-units, as an unsuspected evidence of the utility and application of the knowledge of regional geomorphology. The idea was formally presented by Christian, G.S. (1957) at the 9th Pacific Science Congress, and reproduced in its proceedings. The mapping of the terrain by an inter-disciplinary group, of which geomorphologist was an indispensable member, ensured a broad understanding of the land with its soil and vegetation, usually on a scale of 1 : 250,000. Terrain Classification maps also been prepared in U.K., parts of Africa and the USSR, not necessarily following the landscape methods. In India organised 'Terrain classification' studies have been neglected. What has been emphasised in their place is the 'terrain evaluation' which generally aims at understanding the accessibility and mobility for military manoeuvres. Some studies of this kind do appear for small areas in Ph.D. dissertations,* but I don't know of any programme that aims at an organised effort for systematic terrain classification mapping of India. It would be a worthwhile exercise to submit a proposal, to the funding agencies of the Government of India, to organise terrain classification mapping on 1:250,000 scale for the whole country. Even earlier, though without much participation of geomorphologists but requiring an understanding of geomorphology, were the reports of the Royal Commission on Coast Erosion and Afforestation (1906) not only outlining the level of sedimentation and erosion but also recommending measures for the prevention of the damage caused by these. The U.S. soil conservation service and the Land Research Division of the CSIR, Australis, made extensive use of geomorphology. Bennett (1928) demonstrated the importance of the applicability of geomorphological understanding in soil conservation, while working for the U.S. department of agriculture.

During the past three decades, the geoscientists in general and the geomorphologists in particular, seem deeply concerned about soil erosion, sedimentation, land-degradation, leaching of the soils and their infertility and the damages caused to agricultural land by episodic events like flash foods and duststorms. This is indisputably borne out by the publications on 'Land and Soil Surveys, and planning for land development by geomorphologists such as Jacks and Whyte (1939), Stewart (1968) and Young (1968, 1973). The usefulness of geomorphological knowledge has been easier to demonstrate after the advent of 'Remote Sensing Techniques' which have since been revolutionised. Not confined only to the traditional aerial photographs, which had a smaller coverage and demanded proper weather conditions for aerial surveys, use is made today of satellite imageries produced on the basis of variable electro-magnetic radiation from the heterogeneous earth's surface and the ground cover. Since these are taken from much greater heights, the coverage is more extensive and the quality of resolution reasonable. The Dutch geomorphologists led by Verstappen (1977, 1983) are the leaders in the utilisation of remote

* Some of these studies are mentioned by the authors in their paper. Prudhvi Raju, K.N. *et al.* (1989), "Geographic Information Systems to evaluate land for environmental management". *Annals of NAGI*, Vol. IX 19-27.

sensing techniques. Verstappen advocates the application of geomorphology in the field of earth science (topographic and thematic maps) environmental studies, rural development, urban problems, and engineering.

Some Important Areas of Geomorphic Application

Geomorphology applied to Land and Soil

To my mind this is by far the most significant application of geomorphology and of great relevance to the Indian situation. Besides the mapping of static land-units talked of earlier, that could provide a feel of the land broadly, the processes at work in an area keep the land in a state of flux. Gullying, loss of soil through sheet erosion, deposition of sediments burying the cultivated fertile soils, the slope wash processes, the river bank erosion, land-slides and a variety other secular or episodic events may produce damaging effects that could be stopped short of being catastrophic and in some cases completely eliminated. The recognition of the facets of relief and the breaks in the regional transverse profiles are diagnostic, and the recommended land management practices based on their understanding will help retain the stability of the land. Likewise, the lands subjected to sheetwash, following the convergence of the gullies on the hill slopes, could be saved by a diversionary device. The inundation of the flood plains, however, is not an unusually catastrophic event. In some situations, such as the Nile delta, underflooding as much as the overflooding results in the occurrence of famines. The building of dykes and the rechannelling of the river, to minimise the risks of floods, or the provision of spill-over channels and the inundation canals – all require a geomorphic understanding. In fact, there has developed an entire range of practical applications of geomorphology included under the category 'applied-fluvial geomorphology' (Schumm, 1977).

Soil Genesis and Typology

Though different in meaning and approach, weathering profile is not very dissimilar from a pedogenic profile, and is a forerunner of the latter. The pedogenic processes of leaching, eluviation, illuviation and cheluviation and the development of 'O' horizon at the top operate as a subsequent process on a weathering profile. Weathering studies have a direct application in the determination of the nature and richness of the soils. Varying with the climate and the parent material, the soil forming processes could be intensely rapid or extremely slow, thereby causing a positive or negative pedogenic balance *vis-a-vis* erosive processes. Nowhere the role of geomorphology is as direct as in the genesis of soils and their properties. Unfortunately, not many geomorphologists have taken up the study of pedogenesis, offering explanation in the light of the weathering conditions as well as the existing and inherited weathering profiles. The autochthonous soils, in the tropical areas of rapid weathering, reflect in their chemical composition, reaction, and above all in the ion-exchange capacity, the conditions and the consequences of weathering.

Greater attention, no doubt, is paid to soil erosion and its rates, by direct measurement in the field, or through monitoring a channel cross-section at certain control points, in terms of discharge, sediment load and its characteristics. Hudson calculated denudation rate for the USA (a rate of 3 cm/1000 years.) This is a very generalised picture and may not conform to the rates in different parts of the sub-continent. But an accurately determined sediment yield is of enormous utility in understanding the soil-losses from the catchment and the possibility of silt-deposition in the lower reaches of the stream and the flood plains. The immediacy of the problem concerning the retention of the top-soil, and the fact that the water-erosion claims a major segment of soil losses, justify the geomorphic approach to work intensively on the excretary function of the drainage system to understand the metabolism of landscape and recommend and initiate remedial measures. The identification of the areas of accelerated erosion as a result of man's interference is the first step in that direction.

Hydrograph Analysis River-regime and Flood Control

This is not to make a tall claim but only to suggest that in smaller basins, it is possible to predict the discharge level of following heavy rains in the catchment with reference to the hydrograph which establishes a relation between the precipitation and the discharge. The device to control the flood, like holding the water in reservoirs or diverting it into inundation-canals or even by building of levels is a long-term one. Till then, a timely prediction of flood-based on the intensity and the amount of precipitation could lessen the risk to life and property.

Land-slides and Avalanches

These events usually occur in the cold temperate latitudes as well as in the tropical mountainous areas with heavy rainfall. The factors that generally promote the occurrence of land-slides include a wetting of the sub-surface layers, usually covered with weathered mantle, and the development of a lubricated surface to reduce the internal as well as the inter-stratum cohesion such that the force of gravity exceeds the cohesiveness of the rock-mass, either detached from the understratum or falling loosely under its own weight. Sedimentary formations subjected to the fracture, in a fault zone with increased permeability may be subjected to land-slides. A mobile belt like the Alpine or the Himalayan belt or the one characterised by a steep-scarp will easily produce a land-slide following a slight tremor. Laying of roads to improve accessibility, baring of the slopes, excavations for a variety of purposes, overgrazing and deforestation have all added to catastrophic land-slides. In some countries, and I am familiar with Switzerland, an interdisciplinary team involving engineers, geophysists and geomorphologists tackle the problem of land-slides, demarcating hazards areas, stopping excavations, grazing and even cultivation in some sensitive areas, and laying alternative routes wherever necessary.

Applied Geomorphology and the Problems of Arid Lands

No part of India is better acquainted with the problem of arid landscape then Rajasthan. The main problem of these areas emanate from the mobility of aeolian sand, scarcity of water, flash floods, ravine erosion and above all increasing desertification-physical or climatological. The range of studies made on arid regions is very wide, and almost all deserts of the world have been studied by one group of geomorphologists or the other. Desert plains (McGee, W.J., 1897) and the Sand dunes (Cornish, V., 1897) evoked considerable interest in the closing years of the last century. The Germans led by Passarge, S. (1904) showed deep interest and involvement in the Kalahari desert and subsequently in Sahara desert where they had established a field station at Tibesti.* many of these studies following Davis (1906) were devoted to the changes in the arid landscape. Relatively more

* The field-station at Tibesti established under the leadership of Professor J. Hovermann has since been closed for political reasons.

recent works of Cooke and Warren (1973) and Mabbutt (1977) have paid more attention to the problems of the desert regions. A reflection of this can be seen in Cooke's (1977) 'Applied Geomorphological Studies in Deserts'. Replete with the examples of applied studies on arid lands, this paper is conspicuous by two characteristics, viz., a complete inventory of applied studies, a kind of prospectus, and a table giving the objectives and methods of aeolian sand control, besides a 'summary of deserts regions' after Perrin and Mitchell (1969) and a 'Numerical System of nomenclature' after Grant (1972). The methods suggested may not be valid in all situations but their general merit cannot be disregarded. The latest in the series of attempts by Cookes and his associates (1982) 'Urban Geomorphology in Drylands' is a work that discuss the problems of geomorphic application ranging from salinity, groundwater and salt weathering in drylands to the problems of construction engineering and the problems associated with sand and dust movement. In the Indian context, the Central Arid Zone Research Institute and the University of Rajasthan have been active investigators of arid landscape. Singh's (1977) 'Geomorphological Investigation of Rajasthan Desert' was a prelude to more advanced work in the Institute, where Ghose, Singh, Kar and Pandey did pioneering work on the geomorphology of Rajasthan desert. Sharma (1979) explained the nature and origin of ravines in Chambal basin, a work that could be of immense value in reclaiming or stabilising these ravines. His morphogenetic study of Rajasthan reproduces (1987) from the field a series if pedologic profiles and quaternary sequences which are of value in the management of desert.

Besides Rajasthan, some geomorphologists are active in other regions where application of geomorphological knowledge could benefit the planner engaged in land, soil and water management. The sediment discharge study in Kosi basin by S. Pal and his students, the study of land-slided by Basu in North Bengal and the study of Deltaic areas, old strand lines, left-over and buried channels and the advance of the coasts by R. Vaidyanathan and his students are some of the contemporary attempts by geomorphologists in India. Mention has to be made of the Earth Science Centre at Trivandrum, where some geomorphologists are engaged in coastal monitoring and some are trying to establish a model incorporating facets of relief and landuse in Kerala. There may certainly be several other geomorphologists who are working in the area of applied geomorphology, but, I must plead ignorance. To my understanding, no proper survey is done of 'who is doing what'. It will be the function of this forum to compile an inventory of the themes on which geomorphologists in different parts of the country are working.

Without in the least intending to underestimate the importance of the work being done in India, or even globally, one must admit that the application of geomorphology to the problems of land, soil, water and engineering works has not gained the necessary momentum. Part of the reason lies in the general reluctance on our part and the contempt with which the subject is treated by some internationally distinguished geomorphologists. As a fraternity, we observe and discuss profusely, do a fair amount of writing, adding to the growing mass of literature on the subject, but we are poor practitioners of the science of geomorphology. A fact often overlooked is that a problem-oriented approach may provide the much needed feed-back for sharpening our understanding of the principles.

The Present Day Application

Much of the application of geomorphology today is largely in the field of cartography, production of maps that can demarcate areas susceptible to different kinds of environmental hazards, may these be the maps of areas susceptible to floods, land-slides, various intensities of erosion, gullying and sedimentation, or may these be the areas of wasteland, degraded forests or overgrazed grasslands. Many of these exercises are based on the application of the remote sensing techniques. In India, there is a virtual explosion of the use of imageries on a scale of 1 : 250,000, or their blown-up prints. The soil surveys, the groundwater boards, the forest departments, and almost all University earth-science departments are making use of these imageries, backed as they are also by Space Application Centre of ISRO. There is little doubt that the synoptic view of a large area, and identification of many geomorphic, pedologic or silvicultural features, have made the job of the earth scientists easy, saving time and resources, yet the dynamic aspects of field realities are still inescapably important and cannot be dispensed with. Preparation of thematic maps from imageries is partly geomorphic mapping and partly geomorphic application, but many other dimensions of real application, which require a data-base derived from field and laboratory experiments and measurements, cannot be substituted by imagery based cartographic exercises.

The Present Situation

At the moment, two contradictory trends are discernible in the field, one that emphasises the 'quest for utility' and is committed to transform geomorphology into a geotechnical discipline with all the existing and developing tools at its command, and the other that dreads the idea of reducing 'a geomorphologists to a technician and the discipline to an intellectually barren field. Yet the social pressure make it imperative that the geomorphologists establish the relevance of their discipline. If one has to speak plainly, it is far-fetched to claim a status for geomorphology, comparable to Mathematics, physics and chemistry. The geomorphic principles largely reflect the interaction of physical laws, and to that extent are only a derivative. The subject can be firmly established on an unassailable ground only by recognising the reality of its existence. While both the basis and applied nature of the subject are important and complementary, the welfare of the humankind is the focal point in the development of all discipline and geomorphology is no exception. Should we not, then, practice the art of geomorphology to redeem our existence.

REFERENCES

Bennett, H.H. (1955): *Elements of Soil Conservation*, New York, McGraw-Hill.

Cooke, R.U. and Warren, A (1973): *Geomorphology in Deserts* (Balsford, London).

Cooke, R.U. (1977): 'Applied Geomorphological Studies in Deserts in Hailes John R. (ed.) *Applied Geomorphology Elsevier*, Amsterdam.

Cooke, R.U. and Doornkamp, J.C. (1974): *Geomorphology in Environmental Management*, OUP, Oxford.

Cooke, R.U., (Brunsden, D., Doornkamp, J.C. and Jones, D.K.C. (1982): *Urban Geomorphology in Drylands* Oxford University Press.

Cornish V. (1897): Sand Dunes, *Geog. Journal*, Vol.9. Passarge.

Cristian C.D. (1957): 'The Concept of Land Units and Land Systems' proceedings of IX Pacific Science Congress.

Denys Brunsden (1985): 'Geomorphology in the Service of Man' in Johnston R.J. (ed.) *The Future of Geography* Methuen, London.

Davis, W.M. (1906): 'The Geographical Cycle in an arid climate' *Geog. Journal* Vol. 27.

Grant, K. (1972): Terrain Classification for Engg. Purposes of the Melbourne Area, Victoria, CSIRO, Aust., Div. Geomorph. Paper No.11, pp .209.

Ghose, B., Pandey, S and Kar, A.(1977): "Desertification around the Thar" *Annals of Arid Zone* 16, pp. 290-301.

Ghose, B., Pandey, S and Singh, S. (1968): "Process and Extent of Erosion and its effects on Landuse in the Central Luni Basin , *Annals of Arid Zone*, 7, pp. 15-30.

Hails John H. (ed.) (1977): *Applied Geomorphology* Elsevier Scientific Publishing Co. Amsterdam.

Hart, M.G. (1986): *Geomorphology: Pure and Applied,* George Allen and Unwin pp. 130-132.

Hudson, S. (1968): Erosion of Land, *Am. Scientist* 56:356-74.

Hudson, S. (1971): *Soil conservation,* Batsford, London.

Jacks, G.V. and Whyte, R.O. (1939): *The Rape of the Earth — A World Survey of Soil Erosion,* Faber, London.

Mabbutt, J.A. (1977): *Desert Landforms,* MIT Press, Cambridge, Mass.

McGee W.J. (1897): " Desert Plains" *Bull-Geol. Soc. Am.,* Vol. 8

Ollier, C.D. (1977): "Terrain classification: Methods applications and principles, in Hails, S.R. (ed.) *Applied Geomorphology,* Elsevier, Sci. Pub. Co. Amsterdam.

Perrin, R.M.S. and Mitchell, C.W. (1969): An appraisal of physiographic units for predicting site conditions in arid areas Mil, *Eng. Exp. Establ. Rep.* Vols.1 and 2.

Passarge, S. (1904): *Die Kalahari,* Berlin.

Schumm S.A (1977): 'Applied Fluvial Geomorphology' in Hails John R. (ed.) *Applied Geomorphology.*

Sharma, H.S. (1980): *Ravine Erosion in India,* Concept, New Delhi.

Sharma, H.S. (1987): *Tropical Geomorphology: A Morphogenetic study of Rajasthan,* Concept New Delhi.

Sharma, H.S. (1979): *Physiography of the lower Chambal Valley* Concept, New Delhi.

Singh, S. (1977): Geomorphological Investigation of the Rajasthan desert, Central Arid Research Institute". Jodhpur monograph No.7.

Verstappen, H.Th. (1977): *Remote Sensing in Geomorphology,* Elsevier, Amsterdam.

Verstappen, H.Th. (1983): *Applied Geomorphology: Geomorphological Surveys for Environment development* Elsevier Amsterdam.

Young, A. (1968): Material Resources Survey for Land Development in the Tropics, *Geography,* 53, pp. 229-48.

Young, A. (1973): 'Soil Survey Procedures in Land Development Planning' *Geog. Journal,* 139(1), pp. 53-64.

2

Environmental Geomorphology and Soil Sciences and Their Application to Ravine Reclamation

M.L. DEWAN

I AM pleased to address members of the Indian Institute of Geomorphologists on their Second Annual Session. The theme is "ENVIRONMENTAL GEOMORPHOLOGY". You all know environment has a number of meanings and so also geomorphology. The effect of soil science on environmental geomorphology can also be interpreted in many ways. I will, however, give my key-note address by presenting the various aspects of Ravine Reclamation not only in Rajasthan, but in other parts of India also.

I had the privilege to address the Soil Conservation Society's symposium on Ravine Reclamation at Agra where this Society met two and a half years back (March 1987). It had many specialists who worked on it, some of them for their life-time. There was, however, difference of opinion as to what has been the progress in the past forty years since Independence. Being a Soil Scientist, I am tempted to giving the view-points on soil and ravine reclamation. But that will be limiting it too much. What would be of interest to the present audience will be the genesis of ravines, their impact on environment as well as *vice versa* and what can be done to make optimum use of these land resources. What would be an integrated land and water use in ravine area? How to optimise and integrate issues for the people living in or close to such ravines?

Many of the studies on this subject have been made by the people who are present in this seminar or who are managing this conference. For example, the following publications by Dr. H.S. Sharma on

(i) Ravine erosion in India, (ii) Physiography of the Lower Chambal Valley, (iii) Tropical Geomorphology and (iv) the four volumes on Perspectives in Geomorphology" published by Concept Publishing Company, Delhi, highlight some of the work on ravine formation and reclamation.

The ravine erosion is extremely serious in the four scattered parts of India:

(i) The Yamuna-Chambal Region
(ii) The Sub-Himalayan Zone in Punjab
(iii) The Gujarat plains, and
(iv) The Chhota Nagpur Plateau.

An estimate prepared by the Planning Commission (in 1972), indicates that ravine erosion has damaged 36.6 lakh hectares of land – Uttar Pradesh 12.3 lakhs; Madhya Pradesh 6.8 lakhs; Rajasthan 4.5 lakhs; Gujarat 4.0 lakhs and also in five other States to make a total of 36.7 lakh hectares (the estimate in 1987 was about 40 lakh hectares). The ecological factors of ravine erosion include the geological structure, the climate, rainfall features, the vegetation and geomorphological factors as well as landuse. These and many other subjects such as morphology, ravines-genesis, ravine reclamation, planning and operations are given in the books by Prof. Sharma, and I shall take this opportunity to compliment him and his colleagues for such a good work indicated in these publications. It is, therefore, with some degree of humility that I like to present my views and experiences on ravine reclamation both the principles and practices of it, as I believe, it should be applicable in various parts of Rajasthan, Gujarat, Madhya Pradesh, Maharashtra and Uttar Pradesh. I give below some of these principles and then follow it up with some practices. I could not do better than quote parts of my earlier paper entitled *Reclamation and Greening of Ravines - A Call for Action*:

The National Policy on Ravine Lands

The National Landuse and Wasteland Development Council has taken the following decisions regarding ravine lands:

(i) Special programmes of reclamation and rehabilitation in the ravine areas must be launched with greater vigour in order to prevent wind and water erosion.

(ii) In remote and difficult areas in critical watersheds and in ravines, aerial seeding must be considered.

(iii) People's involvement in nurseries and three growers cooperatives must be promoted.

The time has come for a call for action for launching a well-considered action programme for reclamation and management of ravines through involvement of all sections of people concerned.

As a guess estimate, only 10 per cent of approximately four million hectares of ravine land are under some productive use for wood fuel, pastoral forest, horticulture, crops, etc. So there is a vast potential resources which otherwise can deteriorate fast as evidenced by many studies.

It is proposed that the call for action be on the following lines:

(i) To develop a strategy for reclamation and greening of ravines. Such a strategy, it is proposed, should be based on the following pillars:

(a) Close as much of the area in ravines for pastorages by the cattle as possible. Such a closure should be done through bio-fencing (example Agave) or social fencing where people are trained and advised not to let their cattle in such ravines (alternatives for stall feeding and other means must be made available to them).

(b) Plant as much of the area for permanent vegetation. *Examples* — Acacia which is a very good fuel wood, *Subabul* which is multi-purpose, Prosopis — also fuelwood and other species of fuel, fodder and multi-purpose trees. The tree planting programme will require nurseries as well as seed and plant material, collection and distribution centres in all districts of ravine lands. This is a huge task which will involve people's nurseries, school nurseries, nurseries at the Centres of Research and Development, nurseries at the District Headquarters as well as Block Headquarters.

Water is the life-blood for any plant production. As much of water-harvesting as can be done should be encouraged. The gully head erosion which is so typical of ravines should be stopped at the source so that ravines do not expand in area.

People must form themselves into associations such as Ravine Reclamation Association or Tree Planting Association or other cooperatives whereby land resource in well-defined units are taken as the area for action.

Marking price support, bank loans and other supporting programmes be ensured for the genuine people's cooperatives which must have time-bound programmes and projects for such reclamation and greening as well as utilisation programmes.

A call to the Government were all members of the ravine reclamation projects of the Government as well as Central Soil and Water Conservation Research and Training Institute in Agra (U.P.), in Vasad (Gujarat) and Kota (Rajasthan) should be involved in action programmes through helping the people's associations in planning programmes, giving technical support as well as monitoring and evaluation. Such institutes can also help people's association for obtaining funds from the funding resources.

Reclamation of Ravines are explained in a Call for Action. These are given below.

ENVIRONMENTAL PROJECT PREPARATION FOR PRO[illegible]TION OF LANDS DEGRADED BY RAVINES

Principle: PROTECT AND PRODUCE

Part I

1. **Introduction**
2. **Soils**

2.1 Types
- Red Laterite
- Black Cotton Soils
- Alluvial
- Brown Forest
- Shallow (Skeletal) Soils
- Saline/Alkali Soils.

2.2 Characteristics
- Colour
- Texture
- Structure
- Depth

Chemical Characteristics.

- pH
- Base saturation

2.3 Suitable Land Use

2.4 Modification

3. **Plants**

3.1 Trees (Agro-Forestry)

- Fuel
- Fodder = Quick Growing
- Timber = Slow Growing
- Pulp and Paper - Nitrogen Fixing
- Multi Purpose - Nitrogen Depleting
- Horticulture - Fruit Trees

3.2 Bushes

3.3 Grasses and Legumes

3.4 Combination

Multi-storey plantation Approach

Why Trees - Roots reach where shallow root-system (Nutrients - Water) Plants do not

We are lucky

- Abundant Solar Energy
- Monsoon Land

3.5 Nursery Operations

- Dynamics
- Principles
- Practice
- Seed Collection
- Rhizomes/Plant Material

3.6 Plantation

- Pit Method
- Spacing
- Ditch

3.7 Fencing and After Care

- Bio Fencing
- Social Fencing
- Other Types Normally not economical

- Take Care of Each Plant
- Irrigation
- Manuring

4. Water Harvesting, Storage, Management

- Principles
- Quality
- Quantity
- Additional Catchment
- We are lucky, We have already pits (ponds, gullies, storage sites)
- Pumping (Solar Pumping ?)
- Fishery
- Recreation
- Landscaping

5. Project Preparation

5.0 Objectives

5.1 Pilot Project Area

- Description (Area in ha)
- Survey of present conditions
- Soil Survey
- Geological Survey
- Geomorphological Studies
- Economical Survey
- Topographical Survey (Contours)
- Groundwater Aquifers
- Surface Drainage

5.2 Characteristics of Ravines

- Size
- State of Degradation
- Type of Gully Head
- Active Erosion
- Partially Stabilised
- Stabilised
- Depth (M) Max
Av
Min

- Soil Analysis
- Any seepage during/After monsoons

5.3 Ecological Survey

- Major Vegetative Association
- Chemical and Biological Characteristics
- Initial & potential productivity

5.4 Site Preparation

- Regulations
- Eco-system Approach

5.5 **Reclamation Procedures**

5.5.1 Biological Measures

- List of Plants Species for stabilisation of Ravine areas
- List of ornamental

5.5.2 Water Harvesting

- Principles
- Practices
- Water Management

5.5.3 People's Participation

- Principles
- Practices
- Case Studies

5.5.4 Mechanical Reclamation (Where necessary)

6. **Pilot Project**

6.1 Site Description with map

6.2 Reclamation Process
Integrated Approach

6.3 Cost Involvement

6.4 Benefits

- Social Benefits
- Environmental Benefits
- Economic Benefits

7. **Reclamation of Ravines Degraded Lands is a Challenge**

- Cannot give up

- Costs depend upon methodology
- People involvement
- Degraded lands are an insult to our dignity– We must P & P

8. **Summing-up**

Preparation of up to five
Outline project documents
- ..teers
3 Soil Conservation Project
2 Geomorphologist
Ecologists

References
Reading MAterial
Follow up

Principles and Practices of People's Participation

- Understanding collective action
- Oriented of Action

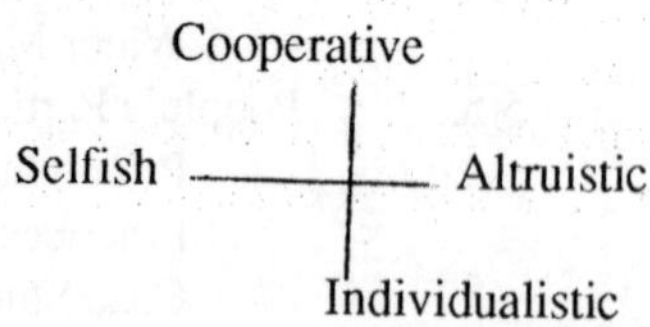

- Managing Common Resources
- Developing Local Institutions
- Bureaucratic Reorientation (B.R.O)
- Promoters for People's Participation
  - Go to the People
  - Live with Them
  - Love Them
  - Learn from Them
  - Work with Them
  - Plan with Them
  - Start with what they know
  - Built on what they have
- Means for Peoples Actions.

To summarise, Science and Technology, People's Participation and

Integration and Inter-disciplinary collective action between Soil and Water Scientists, Foresters, Geomorphologists, Agrotologists, Social Scientists and Economists must go hand in hand to fall in line with Call for Action.

I am sorry I have not given an academic link between environmental geomorphology and soil science in this address because that would have remained in the text-books. What I have presented is my practical experience in Integrated Development of an area of 4 m ha which is the product of Environmental Geomorphology and has unfortunately exceeded the stage of degradation.

Following tools of Science and Technology will be valuable in such Programme of Action–

— Detailed understanding of Geomorphology of such ravine areas and use of Remote Sensing and aerial photography in such understanding
— Rejuvenation and regeneration capacity of such degraded area be strengthened and use of aerial seeding where the degraded areas are not easily accessible
— Involvement of people in such programmes as Government alone cannot *repeat* cannot carry out such stupendous task by themselves. Role of Universities and Academics is valuable in research and investigation and working in pilot projects.

I would, therefore, close by complimenting the University of Rajasthan and especially its Geography Department and Organisers of this conference on such an action-oriented one. I also end with a hope that what I have summarised as my experience in such difficult areas will be worthwhile in guiding and spearheading a Plan of Action.

Thank you for your kind invitation to be with you on this Opening Session.

3

Ecosystems and Environmental Hazards of an Arid Environment – A Case Study

P.C. VATS and SURENDRA SINGH

Introduction

THE AGRICULTURAL progress of an area is affected by several environmental hazards, which are classified as erosional, depositional and salinity. In Barmer district of arid Rajasthan wind erosion/deposition, water erosion, salinity/alkalinity, industrial effluent and mining are the major environmental hazards. Because of these hazards soil productivity decreases, desertification escalates and environmental deterioration fallows. In order to check further degradation of the ecosystem and to improve their agricultural productivity, systematic and reliable information on the type genesis and distribution of fragile ecosystems and associated environmental hazards is of paramount importance. Keeping in view the above, a study was undertaken to map problematic areas for better landuse planning and preventing further deterioration.

Environmental Setting

Barmer district, comprising an area of 28,387 sq km lies in arid zone of western Rajasthan between latitudes 24° 58′ N to 26° 32′ N and longitudes 72° 5′ E to 72° 52′ E.

Climate

The mean annual rainfall of the district is 277 mm. The mean maximum and minimum temperatures in the month of May are 41.9°C and 26.8°C. The mean annual rainfall and number of rainy days are higher in eastern part than the western part.

Material and Method

The study was conducted with the help of aerial photographs of 1: 31,000 scale, landsat black and white transparencies of 1: 1000,000 scale, landsat black and white and false colour composite prints of 1: 250,000 scale and Survey of India toposheets of 1: 250,000 and 1: 50,000 scales were used as basic material for conducting geomorphological and environmental hazards investigations. The field traverses were also carried out along the selected query points to collect the data on the salient geomorphic features, i.e., erosional/depositional, salinity hazards and information on industrial effluent discharged from the textile industries.

Result and Discussion

In the district, ten major ecosystems viz., hills, rocky/gravelly pediments, buried pediments, flat older alluvial plains, sandy undulating older alluvial plains, sand dunes, interdune plains, younger alluvial plains, graded river beds and saline depressions have been identified and mapped (Fig 3.1). These ecosystems were created by the fluvial and aeolian processes under two major wet and dry phases. The ecosystems created by aeolian processes are largely located in the west of 300 mm isophyet, whereas the ecosystems resulted due to fluvial action occur in the east of this isohyet (Fig. 3.2). The salient findings of the investigations are discussed here.

Fluvial Hazards

Interaction of man with the nature enhanced the fluvial geomorphic process which created the sheet, rill, gully, salinity flood and pollution hazards of different intensities. The extent of the area affected under each ecosystem by the above accelerated fluvial hazards viz. water and salinity has been worked out and it is presented in Table 3.1.

(1) *Water Erosion Hazards*

Sheet wash, rills and gullies, created by running water are the major water erosional hazards. These hazards mostly occur on the landforms located in the eastern and central parts of the district.

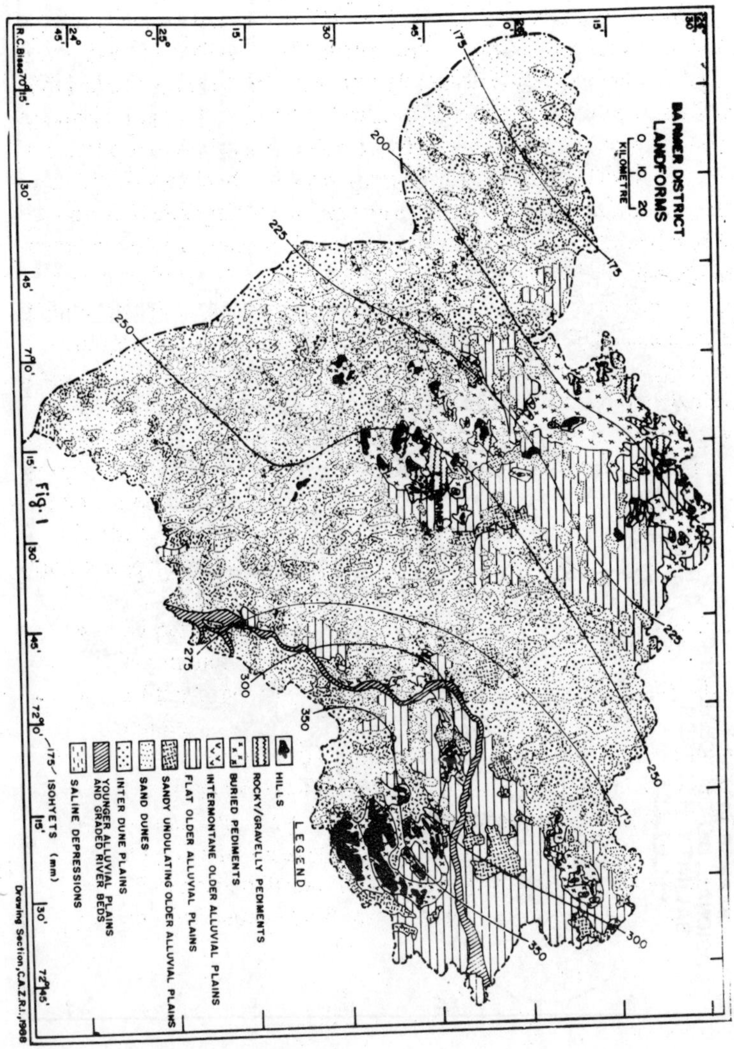
BARMER DISTRICT
LANDFORMS
0 10 20
KILOMETRE
LEGEND
HILLS
ROCKY/GRAVELLY PEDIMENTS
BURIED PEDIMENTS
INTERMONTANE OLDER ALLUVIAL PLAINS
FLAT OLDER ALLUVIAL PLAINS
SANDY UNDULATING OLDER ALLUVIAL PLAINS
SAND DUNES
INTER DUNE PLAINS
YOUNGER ALLUVIAL PLAINS AND GRADED RIVER BEDS
SALINE DEPRESSIONS
175 ISOHYETS (mm)
R.C.Bisa
Drawing Section, C.A.Z.R.I., 1988

Fig. 1

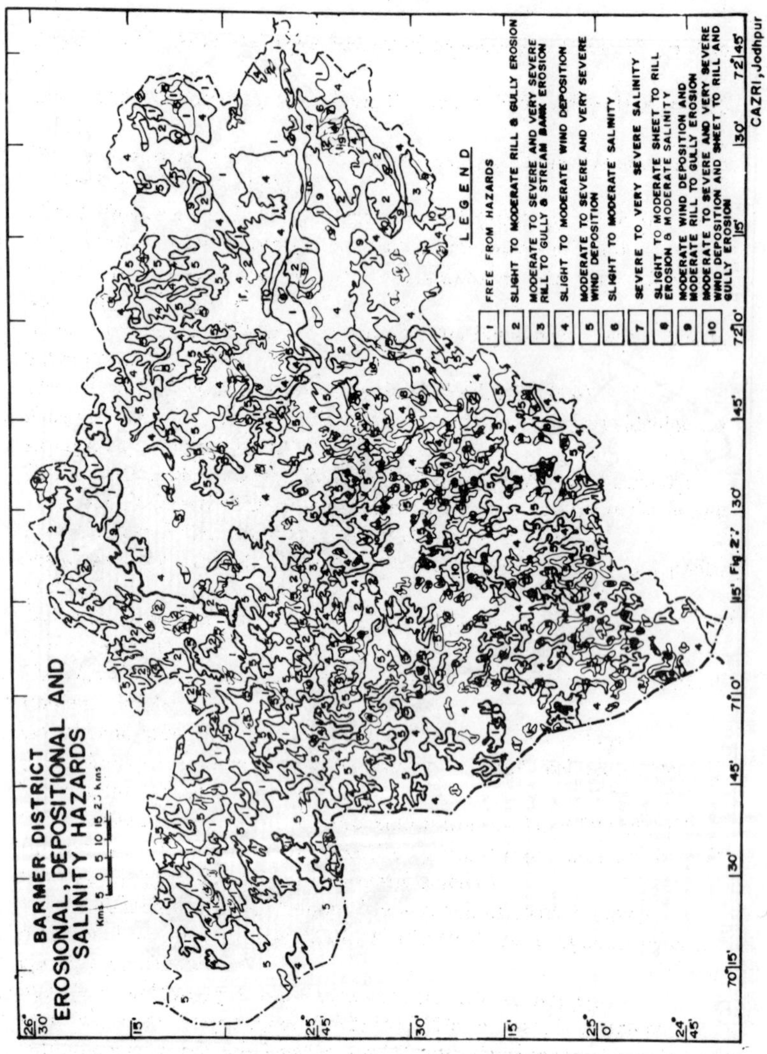

FIG. 3.2: Barmer District: Erosional, Depositional and Salinity Hazards.

TABLE 3.1: Type, Degree and Extent of Erosional/Depositional and Salinity Hazards in Barmer District

Sl. No.	Particulars of hazards	Area in km^2	Percent-age
1.	Free from hazards	5551.37	19.56
2.	Slight to moderate sheet and rill erosion	1124.42	3.96
3.	Moderate sheet to rill and gully erosion	467.13	1.65
4.	Moderate to severe sheet to rill erosion	23.05	0.08
5.	Moderate rill to gully erosion	56.50	0.20
6.	Moderate to severe rill and severe to very severe gully erosion	94.10	0.33
7.	Severe rill to fully erosion	14.45	0.05
8.	Moderate to severe stream bank erosion	1.70	0.01
9.	Slight wind deposition	3839.47	13.52
10.	Slight to moderate wind deposition	4320.32	15.22
11.	Moderate wind deposition	3754.62	13.23
12.	Moderate to severe wind deposition	2202.24	7.76
13.	Severe wind deposition	3613.01	12.73
14.	Very severe wind deposition	1634.53	5.76
15.	Slight to moderate salinity	111.98	0.39
16.	Moderate salinity	40.35	0.14
17.	Severe salinity	16.98	0.06
18.	Severe to very severe salinity	112.11	0.39
19.	Slight to moderate sheet to rill and moderate salinity	52.10	0.18
20	Moderate wind deposition and moderate rill to gully erosion	515.69	1.82
21.	Moderate to severe wind deposition and sheet to rill erosion	316.21	1.11
22.	Severe to very severe wind deposition and gully erosion	524.67	1.85
	Total	28387.00	100.00

Slight to moderate sheet and rill erosion in predominant in the older alluvial plain near Miyon Ki Dhani, Sonklon Ki Dhani and Samdari unscientific cultivation and constant grazing have destroyed the surface stability of this ecosystem. The thin cover of grass and other vegetation has been over exploited which was resulted in the accelerated erosion. Subsequently, surface rain water began to erose the good agricultural land and the kankar pan has exposed to the surface. This hazard has damaged 3.96 per cent area in this district.

Rill and gully erosion hazards (Fig. 3.3) of moderate and severe intensity has degraded 1.93 per cent area of hills/rocky/gravelly pediments and flat buried pediments located in Pachpadra, Sheo and Barmer tahsils near Danta, Mangeriya, Mokalsar and Pusad. The water erosion of severe intensity occurs in 0.38 per cent area of hills, rocky/gravelly pediments and flat buried pediments located in the above tahsils. Because of these hazards large quantity of sediments have been washed away from these landforms resulting into the depletion of biological productivity. The accelerated water erosion has created the rills and gullies of 0.9 to 1.5 m and 1.5 to 3.5 m width and 0.7 to 1.0 m and 1.7 to 2.5 m depth.

FIG. 3.3: Rills and gullies.

Bank cut erosion of the Luni river near Samdari, Majal, Silor, Jethantori, Tilwara and Bamsin during the floods create hazard to the younger alluvial plain of this river. The amount of dissection varied from 3 to 5 m.

(2) *Salinity Hazard*

Another major hazard which limits the agriculture production in Barmer, is the salinity. Natural and man induced salinity has been encountered in the saline depressions, flat older alluvial plains and younger alluvial plains. Slight to moderate man-induced salinity which has come up due to the irrigation from saline water, has affected 0.53 per cent area of the flat older alluvial plains and younger alluvial plains near Rakhi and Karmawas. Slight to moderate sheet to rill and moderate salinity have degraded 0.18 per cent area of older alluvial plains and younger alluvial plains along the Luni river. The salinity has developed as a sequal of flash flood of 1979 near Rakhi, Karmawas, Gadesara and Sara. Severe to very severe natural salinity occupy 0.45 per cent area of saline depressions near Pachpadra, Sanwarla and Thob etc.

(3) *Wind Erosional and Depositional Hazards*

The aeolian activities are mainly responsible for the creation of these hazards in the form of deflation of finer sediments and through the sand deposition. The wind erosional hazard which is generally active in the summer months blow away the top layer of fertile soils of flat older alluvial plains and inter-dune plains, up to 30 to 60 cm depth thus exposing the underlying *kankar* pan and the roots of the trees (Fig. 3.4) near Raslani in Shiv tahsil to the surface. Hollows have been created by the deflation process due to the removal of alluvial and aeolian sediments.

FIG. 3.4: Kankar pan and roots of trees near Raslani in Tahsil Shiv.

Aeolian landforms such as flat and sandy undulating interdune plains, sand dunes and sandy undulating older alluvial plains are subjected to slight to moderate wind deposition hazard. This hazard has affected 41.97 per cent area of these landform units, by depositing fresh sands and reactivating the stabilised dune sands. Moderate to severe and very severe wind deposition occur in 26.25 per cent area of the sand dunes, sandy undulating older alluvial plains and river beds.

(4) *Complex Hazard*

The combined action of wind and water has created the complex hazard of moderate intensity in the form of dune, hummocks, rill and gullies and has affected 1.82 per cent area of the sandy undulating buried pediments, dunes and sandy intermontane alluvial plains near Pachpadra, Shiv and Barmer tahsils. Moderate to severe and very severe complex hazards have degraded 2.96 per cent area of above landforms.

(5) *Pollution Hazard*

The textile industries of Balotra are causing health hazards to the local public. These factories discharge huge amount of liquid waste which include the metals like vanadium and nickel, bleaching powder, caustic soda and sulphuric acid. A huge amount of effluent water of these factories is poured into the Luni river (Fig. 3.5). The effluent water of the industries is highly contaminated (Table 3.2) and remained stagnant and recharge the wells located either sides of the river. As a result the water of the wells of younger alluvial plains near Balotra has become saline/alkaline.

Because of the polluted water the landuse system in the younger alluvial plain of the Luni river has been adversely affected. The water which is used for irrigation contain the anions like the chloride (298 to 1491 mg/l), carbonates (11 to 55 mg/l) and bicarbonates (112 to 495 mg/l) has created the problem of sodicity in the soils (Singh, 1985). The polluted water has also affected the productivity of the land. The yields of wheat from 15 to 22 q/ha has come down from 4-6 q/ha and that of raya from 4 to 6 q/ha has decreased 1 to 2.5 q/ha.

(6) *Mining Hazard*

Mining activities in the district are restricted to a negligible area.

Bentonite near Kapurdi, gypsum near Kawas and Chittar and stone at different places are unscientifically exploited which are causing environmental hazards in the local vicinity, i.e., large pits are left uncared or debris/mining waste are heaped on arable lands. Stocking of gypsum on good agricultural fields lead to loss of biological productivity.

FIG. 3.5: Luni River: Polluted water.

Conclusion

It may be inferred from the study of the ecosystems and environmental hazards that out of the total district area (28,387.00 sq km) 22835 sq km (80.44%) under different landforms is affected by water erosion, wind erosion/deposition and salinity/alkalinity hazards. Remaining 19.56 per cent area is free from such hazards. The wind erosion/deposition is the dominant hazards of the district (Fig. 3.2) and it has degraded 68.22 per cent area of different ecosystems. The complex hazard in the form of wind and water has desertified 4.78 per cent area of the sand dunes and sandy undulating buried pediments and intermontane alluvial plains. Water erosion in the form of sheet, rill and gully occurs 6.28 per cent area of the hills, rocky/gravelly pediments, flat buried pediments, flat intermontane alluvial plains and younger alluvial plains. The salinity/alkalinity hazard has affected 1.16 per cent area of saline depressions, saline flat older and younger alluvial plains.

TABLE 3.2: Results of Chemical Analysis of Grab Samples of Raw Effluents from Some Textile Industries of Pali

Sample Name of of Process(es)	Colour and appearance	pH	Suspended solids (mg/l)	Oil and grease (mg/l)	Chemical Oxygen demand (COD) (mg/l)	Biological demand (mg/l)	Oxygen (BOD)
Desizing	Cream yellow dirty turbid	12.80	1280.0	162.0	760.2	398.0	
Bleaching	Cream yellow turbid	12.50	1185.0	142.0	818.7	547.0	
Dyeing	Deep green turbid	9.85	395.0	36.0	833.3	657.8	
Printing	Bricks red turbid	8.40	235.0	28.0	672.7	474.8	
Finishing	Milky white turbid	7.50	680.0	64.0	710.7	476.4	
Composit effluent (Desizing + Dyeing + Printing	Dirty green turbid	12.50	325.0	81.0	935	571.8	
Caustic treatment + Bleaching + Procion printing	Purple Fluorescent turbid	12.55	495.0	94.0	608.1	383.51	

Source: Mauria, 1986.

Suggestions to Control Environmental Hazards under Different Landforms

Following suggestions for better utilisation of resources and improvement of ecosystems should be considered.

(1) The sand dunes which are geomorphologically more unstable and highly vulnerable to desertification processes should be completely checked from animal grazing and cultivation. The hummocks which are active should be checked by mulching with local plant materials. This should be followed by sowing

grass seeds along mulch lines during rainy season.

(2) To avoid the menance of wind erosion/deposition hazards the deep ploughing in the wind direction during summer season should be discouraged. The ploughing should be done across the prevailing south-west winds.

(3) Controlled grazing should be introduced, if possible, grazing lands should be fenced.

(4) The effluent water of the industries should be released in the river only after the proper treatment and removal of harmful salts.

(5) The irrigation from saline/alkaline water should be stopped and it should be reclaimed by adding gypsum 5-10 ton per ha.

(6) Soil and water conservation measures are needed to check soil erosion and to conserve water resources.

Acknowledgement

Authors are grateful to Dr. J. Venkateswarlu, Director, Shri P.C. Chatterji, Head of Division I, Central Arid Zone Research Institute, Jodhpur for providing facilities to conduct this study.

REFERENCES

Kar, Amal (1978): Geomorphic hazards : Their genesis, nature and problems of identification in the arid zone. Proceedings of the international training course on integrated natural resource survey, pp. 217-222.

Mauria, Shashank (1986): *Textile pollution problem in Pali Marwar.* Desert Environment Conservation and Management (Shankarnarayan and Vinod Shankar ed.) C.A.Z.R.I., pp. 89-93.

Singh, S. (1985): Remote sensing in monitoring desertification process under different landform ecosystems of Indian desert. Recent advances in monitoring the process of desertification, in *Desertification: Monitoring and Control*, Tewari, A.K. (ed.). Scientific Publishers, Jodhpur, pp. 47-73.

Vats, P.C. (1985): Desertification process under different landforms using remote-sensing techniques in Pali district in *Desertification: Monitoring and Control*, Tewari. A.K. (ed.). Scientific Publishers, Jodhpur, pp. 257-264.

4

Geomorphological Aspects of the Processes of Desertification – A Case Study of Sambhar Lake and Environs

H.S. SHARMA

Introduction

DESERTIFICATION IS a term that refers to the process of land degradation that ultimately leads to transformation of productive land into an ecological desert. It is a process involving a continuum change, from no degradation to very severe degradation. The causative agent over the last one or two thousand years or more has been man. The man's misuse of land leads to the degradation of the productivity of vegetation and the fertility of soil as well as to various unfavourable alterations in the soil water balance.

Land degradation by man is initiated when the vegetative cover is reduced or destroyed by grazing, cultivation, mining, road construction, irrigation, urbanisation and other activities that disturb the natural condition. Subsequently, water and wind erosion 'accelerate the degradation process. Finally conditions deteriorate to the point where land is abandoned (Dregne, 1986).

In the past few decades it has become evident that the processes of desertification have been very much enhanced mainly through rapid increase of human and animal population. Recurrent phases of drought in the marginal areas of Rajasthan desert have, of course, accentuated this situation mainly from 1985 to 1988 (Sharma, 1988).

The Study Region

The area lying to the north-eastern part of the Aravalli hills has been chosen for the present study. The detailed study of the area around Sambhar Lake up to Jaipur city has been done for understanding various morphodynamic processes and desertification (Fig. 4.1). This area provides an unique opportunity to study both fluvial and aeolian processes and resultant forms.

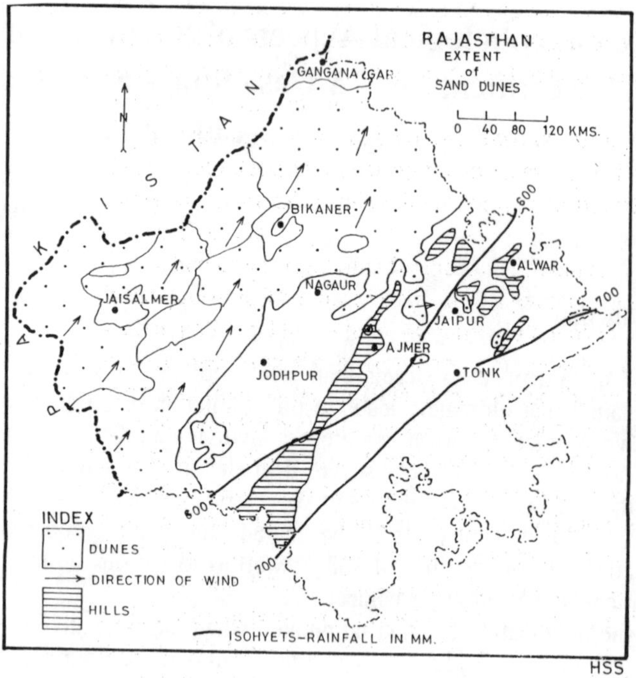

FIG. 4.1: Rajasthan Extent of Sand Dunes.

Climatically, the study region is a part of semi-arid climate. The region receives rainfall between 500 mm and 650 mm and sometimes up to 800 mm. The annual rainfall intensity is worked out as 17.5 mm. The coefficient of rainfall variability is 40.20 per cent. The average annual temperature is 23°C, the maximum being 45°C. The P.E. is 1555.7 mm.

The region under investigation is undergoing gradual desertification mainly due to natural agents (aeolian, fluvial cycles and neo-tectonism), accentuated by anthropogenic activities such as deforestation, overgrazing, unplanned cultivation, expansion of urban settlements, quarrying and engineering constructions.

The Objectives

The main objective of this paper is to elucidate a number of morphodynamic processes enhanced and sometimes triggered off by desertification. The processes discussed in this paper are mainly those of soil erosion, re-activation of stabilised dunes, accumulation of soils through both fluvial and aeolian agents, salinisation, siltation and chocking of river courses in the western part of Jaipur district. It is evident that natural agents become dynamically active once the surface protecting vegetation cover is partially or wholly removed. Such conditions have reaching far negative effects on the eco-potential of the arid and semi- arid areas. In this respect geomorphological studies can contribute to the understanding of manifold problems pertaining to the environment and the methods of its more adapted utilisation.

Methodology

The present study is based on analysis of topographical maps on the scale of R.F. 1 : 50,000, landsat TM data of 1986 and 1987, aerial photographs of 1965 and 1975 on 1 : 25,000 scale with subsequent field checks in different parts of the study region between 1980-1988. In order to ascertain the degree of human intervention on fragile ecosystem analysis of village level data of net sown area has also been done and depicted on map. Aerial photographs have been studied for critical areas where sand accretion, soil erosion and fluvial erosion have been more pronounced.

Geomorphology

The region under study depict an undulating topography consisting NNE-SSW trending discontinuous denudational hills of the Aravallis, isolated hillocks, pediments and pediplains. The intervening valleys between the hills are covered by sand dunes, sand sheets, lake and playa deposits of many cycles. Various climato-geomorphic studies have indicated three distinct phases of aeolian deposits intersected with phases of fluvial deposits in the region around Sambhar lake

(Chattopadhyay, 1978; Roy, 1978; Agrawal, *et al.* 1978; Sharma, 1987).

Geomorphologically, the area can be divided into the following geomorphic units (Fig. 4.2):

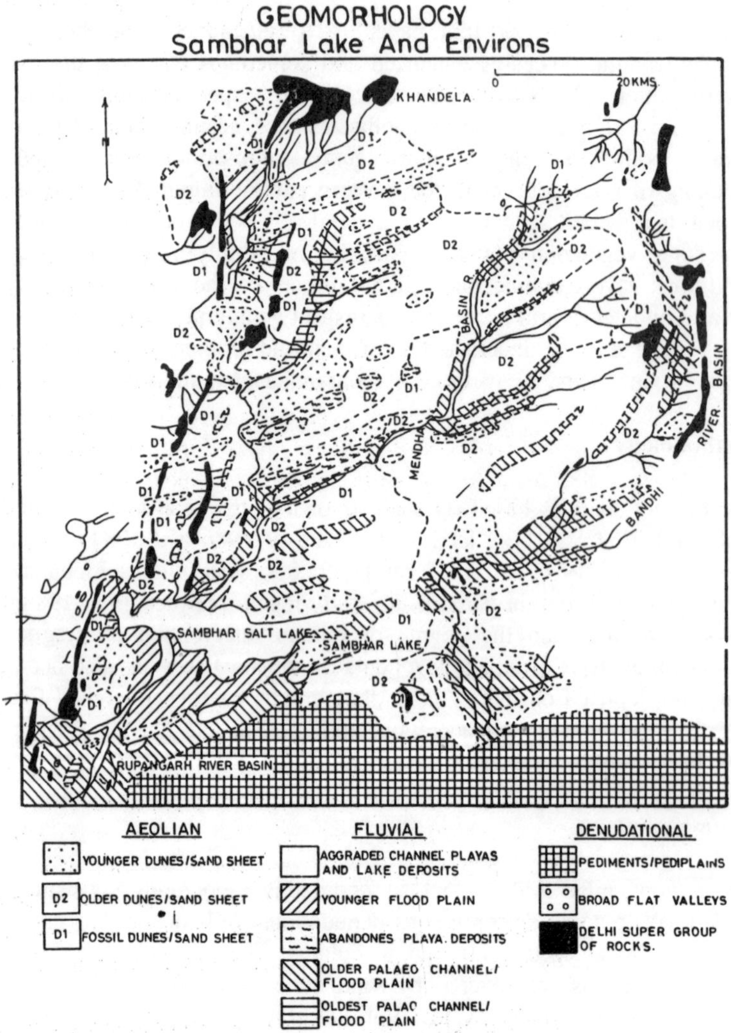

FIG. 4.2: Geomorphology: Sambhar Lake and Environs.

1. Younger Dunes/Sand Sheets

2. Older Dunes/Sand Sheets
3. Fossil Dunes/Sand Sheets
4. Aggraded channel, Playa and Lake Deposits
5. Younger Flood Plains
6. Palaeo-Channels
7. Pediplain
8. Buried Pediments
9. Structural/Denudational hills.

The dunes/sand sheets of the first phase form obstacle dunes around hills and flatter topography in pediplains. The dunes have been subjected to erosion due to gullying and deep dissection (Fig. 4.3). The dunes of second phase are much more stabilised and covered by young aeolian forms. These occur as longitudinal, transverse, parabolic, barchanoid and obstacle shapes. The aeolian forms of the third phase consist of dunes of longitudinal, transverse, parabolic, barchanoid and obstacle types. They are effected by processes of accretion, erosion and stabilisation having different intensity levels depending upon the dominance of the processes operating. In this zone of accretion, these landforms are overlain by small obstacle, linear, barchan and star shaped dunes.

Reactivation of Old Stabilised Dunes/Sand Sheets

A large part of the area to the north-east of the Aravalli hills is covered by old stabilised dunes and sand sheets. A particularly large sand field occurs in the Sambhar area, west of Jaipur and extends almost to Jaipur city (Fig. 4.1). The dunes extend a considerable distance east and north-east of Jaipur. A large number of obstacle dunes are found at Jhir, Banada, Dausa, Bhairana, Jobner, Lalsot, Achroal and Gangapur. This would appear to make the eastern most extension of the sand features in eastern Rajasthan.

The dunal fields of eastern Rajasthan have been misinterpreted as a product of the present aeolian cycle. Geomorphological, archaeological, pedological and botanical evidence suggest that they belong to the extensive belt of old, fixed fossilised dunes of the Rajasthan Desert. These dunes have been formed during the dry phase of the quaternary period. During subsequent more wet phases of climate the eastern dunes have been gradually fixed by vegetation and weathering processes (oxidation or iron and compaction of clay minerals).

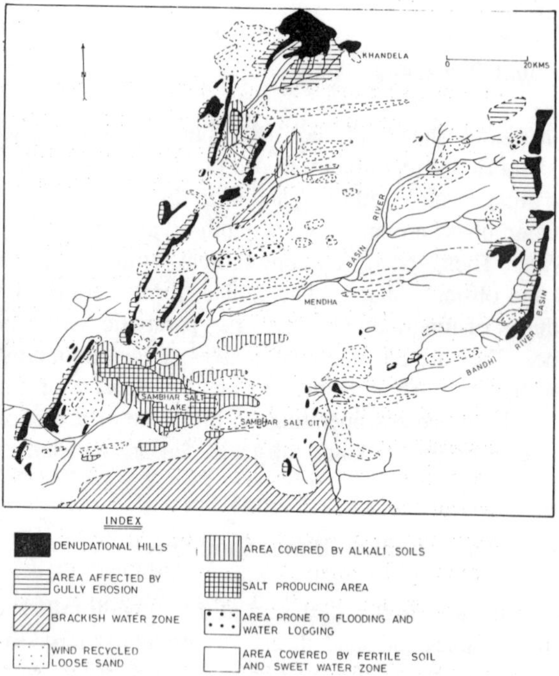

FIG. 4.3: Gemorphic Hazards: Sambhar Lake and Environs.

The field investigation in the Sabi basin indicates the existence of fossil barchan dunes of the bank of the Sabi river. These dunes appear to be the remnants of earlier active barchans which have subsequently been stabilised. These fossil dunes do not retain their characteristic shape owing to subsequent alteration by natural and human agencies. Their existing shape exhibits long parallel sand ridges with cultivated intervening portions. These dunes are also well detected on aerial photographs by their medium gray tones, coarse texture, sparse vegetation and their characteristic topographic expression. Field evidence in Eastern Rajasthan, specially around Jaipur and Sambhar area, make it clear that recent reactivation of these stabilised belts following the rapidly progressing ecological deterioration usually begins patch – like within the affected zone. Further these patches may coalesce to form larger areas. This situation is clearly observed around the large settlements and towns such as Jaipur, Sambhar, Jobner, Chomu and Danta Ramgarh, and depending upon available evidence, one may identify two distinct stages which differ only in degree of anthropogenic dune reactivation in Rajasthan (Fig. 4.4):

1. After a complete removal of the vegetation cover around villages mainly by excessive tree-cutting, overgrazing and over cultivation (Fig. 4.4) such patches appear on the TM imageries and air-photos as blurred, bright patches, more or less star shaped.
2. In the case of more intensified land use the above-mentioned patches marges into large areas. Aerial photographs and satellite imagery show a clear pattern of ripple and barchan dunes which have developed in such areas. In certain cases, where ecological balance has been completely upset, longitudinal dunes develop out of barchans. The dunes show a conspicuous NE-SW and W-E alignment (Fig. 4.4).

More advanced stages of dunes reactivation are pronouncedly marked around Jaipur urban complex, due to the severe interference of man in a very frazile natural ecosystem.

Recent Aeolian Accumulation as an Indicator of Desertification

Observations on recent morphodynamics in north-eastern Rajasthan provide strong evidence to support to the assumption that aeolian accumulation has been enhanced in this century. Sand accumulation has been observed on pediments and slopes of many ridges, inselbergs and other rock outcrops. Heaping of sand on their leeward side is a characteristic feature around Jaipur and almost all denudational hills of the Aravallis in the region under investigation (Figs. 4.3 & 4.4). In more advance stages accumulation of sand could lead to the complete burial of small topographic ridges and old settlements.

Sand Encroachment on Permanent Settlements

Especially in Jaipur and Jobner area, the encroachment of sand masses upon the cultivated fields and their infrastructure has created a dramatic situation. There are evidences in Jhalana area of Jaipur City where buildings were buried under sand encroachment. In Jaipur area only after the recent excessive deforestation, quarrying activity and levelling of old obstacle dunes started to become mobile again and so encroaches upon buildings, road side fencing and chocking of roads during summers and rainy seasons. Since the creation of JDA in 1980 and consistent efforts of Forest Department a programme of fixing these dunes was started partly by conservation and partly by afforestation.

Alteration of the Drainage System

Most of the rivers draining the area are ephemeral of the Kantly, Sobawati, Mendha, Rupengarh have internal drainage where as only Bandi and Dhundh rivers form part of a well-drained system which belongs to the Yamuna network. The present drainage system shows a tendency of headward erosion forming deep gullies in the upper reaches. The river valleys are gradually widening, flattening and aggrading in their lower reaches.

Around Sambhar in the Mendha Catchment wind blown sand has buried many of the seasonal lines during past few decades partly by natural aeolian processes or enhanced by man's interference through over use. The analysis of landsat data and aerial photograph reveal palaeo-channel net under the veil of recent wind blown sand sheets (Fig. 4.4). The drainage of many tributaries is being slowly impeded through the accumulation of sand (the Bandi as an example). The Mendha river bed is gradually being chocked up from Nangal upstream. The width of river has become very narrow and the river bed is completely blanketed by recent aeolian deposits.

Land misuse and over-use and subsequent degradation in the catchment areas of valleys in various parts of the study region have led to drastic alterations in the drainage system. In Jaipur city disturbing well-stabilised dunes in several parts has aggravated gullying as well as chocking of natural water flow and changes in drainage which has resulted into an advanced bandland topography near the city during unprecedented rainfall of July 1981 (Sharma, 1984). The interference with natural drainage through the dunes has meant that run-off, specially during flesh floods, has been arrested. Accumulated water undercut the loose sand entailing foundation problems in building and road constructions with increased run-off, reduced vegetal cover and loosened sand, there is greater erosion than ever before. This was seen earlier in 1978 monsoon and more spectacularly so in July 1981.

Fluvial Soil Erosion

The role of fluvial erosion is pronounced in this semi-arid tract. In fact field evidence speaks of the very strong interrelation of these two agents in this semi-arid area. This is exhibited by the seasonal alteration of fluvial and aeolian morphodynamics within the same area. For instance loose sands which have been transported during the rainy season are transformed into wind blown sand sheets during the dry season.

Areas exposed to this hazards of fluvial soil erosion in the region are mainly the denudational hills of the Aravallis and their pediments. Sandy soils are characterised in poor run-off. But in spite of that, characteristics of tropical areas, can lead to disastrous results that sometimes deep gullies can nag their way backward into vegetation free areas. This is clearly observable around Sambhar, Bhairana, Jobner and Jaipur.

Stronger fluvial erosion takes place on the slopes of pediments. The pediment slopes of the Aravallis have suffered a drastic degradation of their natural vegetation in the past few decades. Lack of vegetation enhanced very much the amount and velocity of run-off, so that most of the top soil has been washed away and severe erosion and gullying takes place on such slopes. The net result may be the establishment of a skaletic soil which are now only used for grazing in the dry season.

Soil Salinisation

The dry surfaces of playa and Sambhar lake show glistening carbonate salts or alkaline soils. Most of the area is lying unused except Sambhar salt lake and Zeen Mata lake where salt is produced. Salt encrustations are widely distributed around Sambhar lake and south east of Danta Ramgarh. These areas are prone to wind blown sand and soil hazards (Figs. 4.3 & 4.4).

Conclusion

It is obvious that the geomorphological processes discussed in this paper can also take place independent of desertification, i.e., under naturally controlled ecosystems, and also in sub-humid and humid areas. Anthropogenically controlled morphodynamics are so interwoven with the naturally controlled ones that it becomes difficult to differentiate between them. Only by a comparison of the processes operating in excessively used areas and in those occurring in underused ones within the same climatic conditions, one can identify and assess the human role in triggering off and accelerating such processes. The present author's attempt made in this direction is rather tentative. It still requires a more detailed presentation of the problem.

REFERENCES

Agrawal, D.P. Dhir, R.P., Krishnamurthy, Mishra, V.N. *et al.* (1978): Multiple evidence for climatic change in Rajasthan in H.S. Mann (ed.) *Arid Zone Research and Development*, Scientific Publishers, Jodhpur, pp. 1-18.

Chattopadhyay, G.S. (1978): Geomorphological Review of a part of Mendha river catchment with special emphasis on desertification problem, in H.S. Mann (ed.) *Arid Zone Research and Development*, Scientific Publishers, Jodhpur, pp. 31-34.

Dregne, H.E. (1986): Magnitude and spread of the Desertification Process, *Arid Land Development and The Combat Against Desertification : An Integrated Approach*, USSR Commission on UNEP, Moscow.

Sharma, H.S. (1984): Impact of unprecedented Rainfall on the Jaipur city and its Environ (Part II), *Annals of the National Association of Geographers, India* vol. iv(i) pp. 4-9.

Sharma, H.S. (1987): *Tropical Geomorphology, A Morphogenetic study of Rajasthan*, Concept Publishing Company, New Delhi.

Sharma, H.S. (1988): The Problem of Desertification and strategy for combating Desertification with reference to Rajasthan, *Indo-Soviet Seminar on Rational utilisation of Natural Resources and Regional Development*, Oct. 10-12, JNU, New Delhi.

Roy, P.K. (1978): Palaeo-climatic aspects in the evolution of the sambhar salt lake - A Geomorphological review, in H.S. Mann (ed.) *Arid Zone Research and Development*, Scientific Publishers, Jodhpur pp. 27-30.

5

Land Capability and Land Management in Kotadun: A Case Study of Central Himalaya — An Ecological Approach

R.K. RAI and R.C. JOSHI

Introduction

THE DIVERSIFIED topography, tectonic unstability, seismicity and climatic variations combined with human interferences, have made Kotadun region fragile and hazards-prone. In the past three decades in the name of regional development for economic growth, more drastic damage has been done creating an ecological imbalance such as accelerating erosions, land-sliding, masswasting flooding, siltation and sedimentation etc., indisciplined and shortsighted human activities such as ruthless exploitation of forest resources, overgrazing, unplanned cultivation, uncontrolled quarrying and construction etc., consequent upon this continuously increasing pressure of human as well as animal population have caused severe land denudation affecting the ecological balance.

The main aim of study is to observe the changing area under major landuse categories on different terrain types and to suggest possible optimal use for each category land.

On the basis of altitude, average slope, geology and morphological characteristics (land-slides, soil erosion, slumping, terraces, gorges and deforestation) study area is divided in three different strata (Table 5.1) The existing landuse has been identified on the basis of land record office data. For getting detail observation one sample village is selected from each terrain. In the last the land of sample villages are divided in different category land taking some parameter (soil, slope, erosion intensity, water holding capacity, sunlight situation, irrigation condition, accessibility from settlement) for the suggestion of proper utilisation of land.

Study Area

Kotadun is the study area of present investigation. Extending in a general east-west direction, this is one of the five duns (Dehra, Kothari, Chaumhamba, Patali and Kota) in the Kumaun Himalaya. Along with its well demarketed boundaries of watersheds of Kumjakharak-Binayak range of Lesser Himalaya to the north and Babilia-Kailkhura range of Siwaliks to the south Kotadun like other duns, is an elongated valley between river Kosi in the west and river Baur in the east. West of Nainital area, it extends between 20°17′39" and 29°30′40" N. latitude, and 79°9′13" and 79°26′24"E. longitude and encompasses an area of 360 km² and a population of 20,657 persons in 1981. Including three forest localities, there are 96 villages.

Kotadun is the combination of two words viz. Kota+Dun. Kota refers to 'Kotha' i.e., place of Chand rulers the ruins of which lie near Ranikota, it was perhaps the winter resort of the Chand rulers. 'Dun' refers to longitudinal structural valley lying between the Siwalik and the Outer Himalayan ranges. In its geological epoch, the outer Himalayan ranges (Upper Siwalik, Dun floor and lower Siwalik) is younger in its stratification and upliftment and is still uplifting at a higher rate then the Lesser Himalaya. The MBF (Main Boundary Fault) occurring between the outer and the Lesser Himalaya is an unstable and fragile zone, and if such a sensitive area is subjected to unplanned or indiscriminate biotic pressure, it becomes highly vulnerable to erosion, masswasting and land-slides. In Kotadun this thrust zone really very hazardous and creates many problems in monsoon season.

In general, the altitude ranges from 400m to 2600m above m.s.l. while some of the peaks of Kumjakharak-Binayak range rise to more then 2600m and those of Babilia-Kailkhura range to above 750m at the point where stream Bangajala terminates in river Kosi, the altitude dips to a little above 400m.

The area is drained four streams system viz. Bangajala, Masani, Dabka and Baur which take their rise in the Lesser Himalaya and drain antecedently through with a gradual decent towards dunfloor furthermore, a gradual rise towards Lower Siwaliks. The Outer

Himalaya is composed of accumulated and uplifted recent debris carried down by the streams from the Lesser Himalaya. Coarse deposition comprising boulders, gravel and sand tends and streams passing through and dun, to drain subterraneously during the lean season. The Outer Himalayan formation is suitable for extensively dense tropical-subtropical vegetation. North of the MBF, on the other hand the gradient increases suddenly become of abrupt rise in the altitude. Steep slopes and crests of the Lesser Himalaya are un-inhavitable and are covered with temperate vegetation.

Most of the settlements in the area occur near MBF on foothill or alluvial fans, in the vicinity of some perennial water source in the Lesser Himalaya. In the Dun floor, these have grown along the large streams viz. Kakurar (feeder of Bangajala), Kaligad (feeder of Masani), Dabka and Baur, so that the water tapped at the foothills, is enough to meet the requirements.

Terrain wise Landuse

Level Area: It comprises an area of 118km^2 or 52 per cent of the total area. 72.45 per cent of this terrain is under control of Forest Department and remaining 27.55 per cent are comprising 35 villages. is governed by Revenue Department. Which major landuse categories indicate 3.8, 0.35, 16 and 72 per cent area under forest, area not available for cultivation, cultivable area and cultivated area.

Ranging between 530m and 1140m the absolute relief is 658m/km^2. Towards southern and northern part dissimilarities of relief are observed but the mid part is almost level. The relative relief ranging between 20 and 340m indicates susceptibility to erosion owing to loose lithological traits. The mean value of relative relief is 56.86 per km^2. Average range of slope is between level to 17°. Here the slope increases with the increasing order of height but it covers lesser area. The average slope is 5.3°/km^2 (Table 5.1).

Mid Slope Area: It comprises an area of 140.8 km^2 or 39 per cent of the total area. Total 75.7 per cent area is under Revenue Department. Total number of villages lying there in is 43. Major landuse categories indicate 18.4, 5.5, 56.4, and 20 per cent under forest, area not available for cultivation, cultivable and cultivated area respectively.

The mean value of absolute relief ranging between 860 and 1700m is 1234m. Relative relief with an average of which is 28°.

TABLE 5.1: Terrain Characteristics

	Level Area	Mid Slope Area	Upland
Altitude	Below 1200m	1200-1700m	Above 1700m
Slope	Below 17°	17°-40°	23°-42°
Geology	Recent deposition Sand and Pebbles	Crushed and brittle rocks	Dolomite, quartzite shale and slate
Forest	Tropical	Mixed decidous	Oak and pine
Morphological Characteristics	Level area with wide river valley and terraces	Landsliding gorges and slumping	Steep slopes and sharp peaks

Upland Area: It comprises an area of 102 km^2 or 28 per cent of the total area. Forest department owns 48 per cent area of this terrain while 52 per cent area of it is under Revenue Department of 18 villages. Major landuse categories include the area of 5.3, 4, 75 and 16 per cent under the forest, area not available for cultivation, cultivable area and cultivated area respectively.

The absolute relief with a mean value of 1748m ranges between 1400 and 2080m. Relative relief ranges between 300 and 580m with an average of 444m. Average slope ranges between 23° and 40° with an average of 31°.

Land Capability

The capacity of land to produce under certain degree of adjustment without deterioration of the soil is defined by Archer (1956) as land capability. Going through the literature (Stamp, 1960; Pant, 1975, 1976, Agriculture Department of U.S.A., 1960) some parameters has been selected for random and easy field-work in hilly terrain is shown in Table 5.2.

Eight variables for each quality land to determine the capability of a particular land or field have been taken in the first phase. The category is determined in following order: (1) Maximum of 150 scode points are allocated to geed quality land and thus each of the variables gets a maximum of 18.75 points in this category. (2) In the same manner the medium quality land of allocated a maximum of 100 points with 12.45 points for each of the variable. (3) Only 50 points are allocated to the poor quality land. Each of the parameter in this class thus, has 6.25 points.

TABLE 5.2: Kotadun: Land Capability Classification Model

Land Capability class	Soil depth	Soil Texture	Natural slope	Erosion intensity	Water holding capability	Sun light situation	Irrigation condition	Accessibility from settlement
Good land	deep 30cm	Loam	Low (10)	Low	High	Sunny	Full	Low (1/2 km)
Medium land	Moderate (15-30cm)	Graval	Moderate (10-20)	Moderate	Moderate	Partially	Partially	Medium (1/2 to 1km)
Poor land	Shallow 15cm	Stony	High (20)	High	Low	Shaddy	Nil	High (1 km)

After Pant et al. (1988).

Applying the land capability model a detail field survey was conducted through the cadestrial maps of three sample villages with the help of planimeter area of different patches was calculated (Table 5.3).

TABLE 5.3: Area (%) under Different Land Classes

Land Category	Villages		
	Patkot	Amgarhi	Sighari
Good	32.97	14.56	-
Medium	61.99	17.56	14.70
Poor	5.05	67.88	85.30

Table 5.3 express level area village contains 33 per cent area under good class. In Sighari (upland village) has no good category land. Medium category land is also decreasing towards upland from level area village. In upland village poor category land is 85 per cent of total village area.

Suggestion and Conclusion

Good land must be utilised for multi-cropping pattern producing food-grains and cash crops. Medium category land should be under double cropping pattern. Keeping in view the scarcity of good quality land non-agricultural uses viz. roads, markets, schools and settlement must be develope in the medium to poor category land. Poor category land should be under social forestry programme to fulfil the requirements of fuel, fodder, timber and fruits. In poor category land which is very hazardous should be under reserve land.

REFERENCES

Pant, S.D. (1935): *Social Economy of Himalayas*, London.

Pant, B.R., R.C. Joshi and D.S. Jalal (1988): Land capability classification model for Kotadun, Kumaun Himalaya, *Geographical Review of India*. Calcutta, 50(i) pp. 47-52.

Stamp, L.D. (1960): *Our developing world*. London.

United States Department of Agriculture (1960): *Comprehensive System of Soil Classification*. Washington, D.C.

6

Land Evaluation and Associated Landuse Systems in Semi-arid Morel River Basin (Rajasthan)

R.D. DOI

THE PRESENT paper makes an attempt to decipher the inherent processes and resultant pattern of land systems and landuse found in a fluvially organised landscape in semi-arid region. The land system approach furnishes a scientific method of land evaluation and its mapping. In fact, land is the stage where the play of man-environmental relationship is witnessed both in time and space. Further, the landuse systems are investigated within the land systems frame to examine the association between the two.

The Morel river basin is located to the east of the Aravalli range between the latitudes 26°6′ 27"–27°9′28" North and longitudes 75°44′, 19°-76°58′ East encompassing an area about 5921 square kilometres. The basin exhibits the superimposition of aeolian landforms over the fluvial landscape.

The study region has been divided into four major land systems and further into units (Fig. 6.1). The investigation is based upon conventional and remotely-sensed data supported extensive random field checks. It could be well established that visual interpretation of RS data help grately in land system mapping of semi-arid landscape. The landuse systems could be evolved after analysing the three year average (1982-85) landuse data for 316 patwar circles in the study region. A very close associations have been found among the various land and landuse systems.

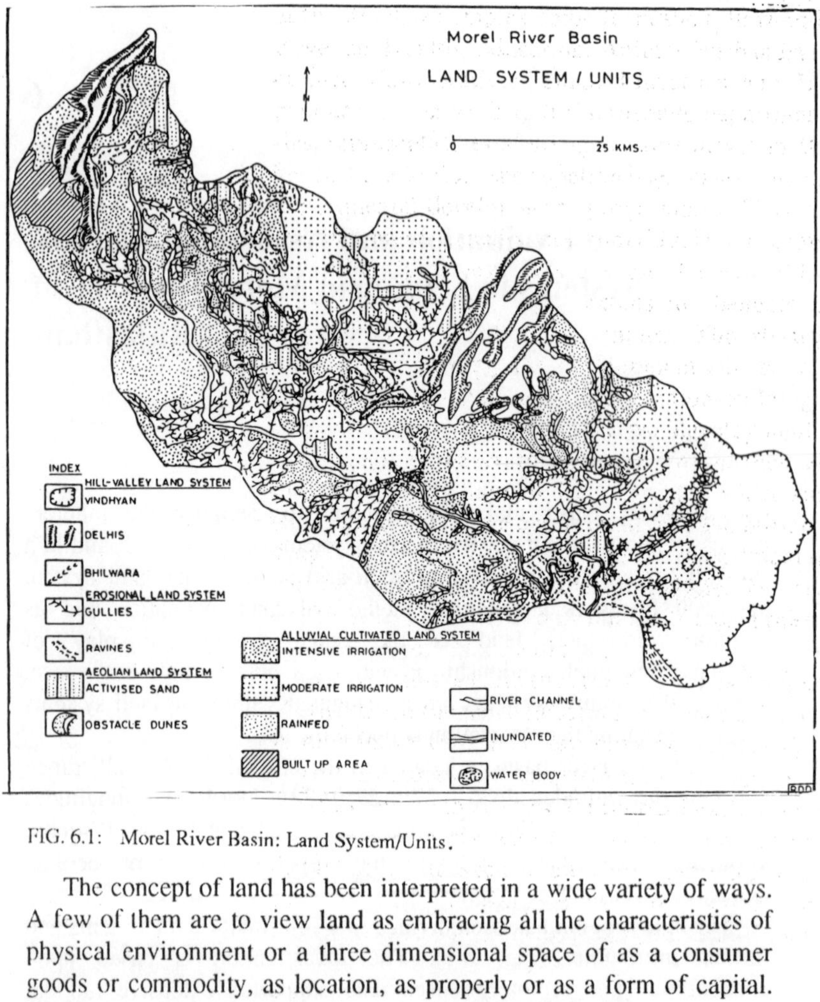

FIG. 6.1: Morel River Basin: Land System/Units.

The concept of land has been interpreted in a wide variety of ways. A few of them are to view land as embracing all the characteristics of physical environment or a three dimensional space of as a consumer goods or commodity, as location, as properly or as a form of capital. Besides, an ecological view is also possible whereby land is equated with nature and is thus defined in terms of ecosystems (Davidson, 1980).

There are various methods of land evaluation. The FAO framework for land evaluation (1976) recommends that land units should be evaluated for land uses in terms of land qualities. The American method of land capability classification is known as the USDA method. Summaries of this method are given by Hudson (1971), Young (1973, 1976), Olson (1974) and Brady (1974). Integral to the assessment procedure is an evaluation of soil erosion hazards, wetness and soil limitation under arid climate. In this scheme land has been divided into eight classes. The classes I to IV are cultivable while the remaining classes are put under unculturable wastes. The Canadian capability scheme has seven classes in contrast to the eight of the USDA method. Like the Canadian method, the British one has also seven classes.

For application of the land system map as a basis for land evaluation, equal attention has been given to recurring pattern of geology, soils, vegetation, geomorphic features, slopes, stream frequency, drainage density and landuse. Use of landset images helps in delineation of land systems which are visible on such images. In fact the terrain governs the distribution of other factors. 'The land systems and landuse types are important indicators of land capability (Doi, 1987).

A land system has been defined as 'an area of group of areas throughout which is recurring pattern of topography, soils and vegetation can be recognised (Christian and Stewart, 1953)'. The main types of land systems/units have been shown in Fig. 6.1. The general characteristics of the land systems/units have been summarised in Tables 6.1 and 6.2.

The associated landuse systems have been found by applying successive quotient method (Kostrowicki, 1976). The ten generalised land systems/units are distributed over twenty-two locations in the Basin, which are listed in Table 6.3.

The landuse associations found within the land systems/units (Fig. 6.2) have been discussed in the following paras:

Hill Valley Land System (HVLS)

In this system, the Vindhyan, Delhis and Bhilwara units are included. In this system the landuse orientation is mainly towards uncultivated land having preponderant to dominating role with forests as far as more important use. In Delhi units pastures shared the largest percentage area of the major category of cultivable land, but in Vindhyan unit the pastures are negligible. The net area sown is less significant mainly unirrigated except in Delhi unit around Amber town where well and canal irrigation is found (Fig. 6.2).

Erosional Land System (ELS)

This system covers gullied and ravinous land units. The landuse is mainly oriented towards unirrigated cultivated land. However, the uncultivated and cultivable categories possess accompanying to preponderant roles with major stress on barren and pastures respectively (Fig. 6.2). The land which was once a continuous flat land with rich soils is gradually becoming irregular by enhanced gully erosion specially after the unprecedented rainfall of July 1981. This fact was established after a good number of enquiries by the author during his field study.

TABLE 6.1: Morel River Basin : Land Systems/Units — General Characteristics

Characteristics	I Hill-Valley Land System (HVLS)		II Erosional Land System (ELS)		
	Vindhayan (V)	Delhis (D)	Bhilwara (B)	Gullies (G)	Ravines (R)
Area (km^2) % to total Basin area	721 (12.2)	445 (7.5)	15.3 (0.25)	1382 (23.3)	347 (5.8)
Geology	Bhander/ Rewa/ Kaimur	Ajabgarh/ Alwar Group	Pre-Aravalli	Alluvium/ Blown Sand	Recent Alluvium
Soils	Alluvium of recent origin/ Mountainous	Alfisol	Mountainous	Alluviums of recent origin to yellowish and grayish Brown	Alfisol
Vegetation	Subsistence/ Edaphic type	Mixed Salar & Anogeissue Pendule type	Hilly/Rocky/ Degraded type	Subsidiary Edaphic type	Subsidiary Edaphic type/ Raverine vegetation
Geomophic	Dissected Plateau	Hog-backs, Flat Tops	Hilly Range	Badland-Topography	Highly Badland-topography
Slope	4°-8° Max.12°	8°-12° Max.20°	4°-12°	Less than 4°	Less than 4°
Stream Frequency	2-9 streams/ km^2	6-8	2-4	4-8	2-4
Drainage Density	2-3 km/ km^2	2-4	1-2	1-4	1-2
Landuse	Forest/ Culturable waste pastures valley fills intensitively cultivated	Hills forest blanks and valley with cultivation	Highly degraded forests/ blanks	Cultivation/ culturable waste/pasture	Barran/ culturable waste

TABLE 6.2: Morel River Basin : Land Systems/Units General Characteristics

Character-istics	III Abolian land system (ALS)		IV Alluvial Cultivated Land System (ACLS)		
	Activised (AS)	Obstacles dunes (O)	Intensive Irrigation (II)	Modrate Irrigation (MI)	Rain fed (RF)
Area (km^2) % to total basin area	294 (5.8)	104 (1.7)	908 (15.3)	1077 (18.2)	791 (13.3)
Geology	Blown Sand/ Alluvium	Blown Sand/ Alluvium	Alluvium	Alluvium	Alluvium
Soils	Loamy sand to sand	Sandy loam	Alluvium of recent origin to yellowish brown and grayish-brown	Alluvium of recent orgin	Alluvium of recent origin
Vegetation	Semi-arid type	Semi-arid	Anogeissue Pendula type and Subsidiary Edaphic type	Subsidiary Edaphic type	subsidiary Edaphic type
Geomorphic features	Sand sheets	Sand falls	Monotonous plain with packed-hills	Monotonous to Undulating plain	Undulating Plain
Slope	Less 4°	4°-8°	Less than 4°	Less than 4°	Less than 4°
Stream frequancy	Less than 2 streams/ km^2	4-6	Less than 2	Less than 2	Less than 2
Drainage Density	1-2 km/ km^2	Less than 1	Less than 1	Less than 1	2-3
Landuse	Patchy cultivation	Barren land	Cropped area	Cropped area	Cropped area

Aeolian Land System (ALS)

Only two Patwar circles of Jamwa Ramgarh tahsil along the northern flanks of the basin are included. This system is characterised by the activised sand dunes and blow sand. The cultivated land is having preponderant role but patchy in pattern with irrigated farming. Irrigation is done with wells along the stream bank where the ground water condition is good. The uncultivated land is dominated by forests while the old fallow land possesses considerable proportion of cultivable land (Fig. 6.2). The ploughing up of sand dune slopes has activised the sand. Consequently it is spread over the nearby fields thereby turning them into old fallow.

TABLE 6.3: Moral River Basin : Land System/Units with Location Code Number

	Land system		Units	Location Code Number
I.	Hill-Valley Land System (HVLS)	(a)	Vindhyan (V)	1
		(b)	Delhis (D)	2 & 3
		(c)	Bhilwara (B)	Omitted
II.	Erosional Land System (ELS)	(a)	Gullies (G)	4 to 8
		(b)	Ravines (R)	9 & 10
III.	Aeolian Land System (ALS)	(a)	Activised Sand (AS)	11
		(b)	Obstacle dunes (O)	Omitted
IV.	Alluvial Cultivated Land System (ACLS)	(a)	Intensive Irrigation (II)	17 to 20
		(b)	Moderate Irrigation (MI)	
		(c)	Rainfed (RF)	21 & 22

Alluvial Cultivated Land System (ACLS)

The land under this system comes nearly 50.0 per cent of the total area of the basin mainly located in a vast tract in the central parts. However, patches of considerable size are found all over the basin. In this system the land is oriented towards cultivation. The net area sown possess dominating to predominating roles. The cultivated land is mainly irrigated ranging from moderate to intensive irrigation, except in western patches where rainfed cultivation is done due to sandy flats and low rainfall. The cultivable land holds the accompanying role (insignificant) mainly oriented towards pastures followed by current fallow. The uncultivated land is negligible dominated by non-agricultural uses and at places accompanied by barren land. In two units only the uncultivated land possesses equal to preponderant roles.

Conclusion

Landsat images help greatly in land system mapping. However, the landform can be determined in air photos more objectively. It is clear that there is no general prescription for the legend hierarchy in land system mapping. Two legend levels have been utilised for the land system mapping in this study.

FIG. 6.2: Morel River Basin: Land Systems and Associated Landuse Systems.

In the Hill valley land system, landuse orientation is mainly towards uncultivated land having preponderant to dominating role with forests as important use. In Delhis units pastures share the largest percentage area of the major category of cultivable land. In Erosional land system the landuse is mainly oriented towards rainfed cultivation. The Aeolian land system is characterised by cultivated land, which is having preponderant role. Finally, the Alluvial cultivated land system covers nearly 50 per cent of the total Morel river basin, which is mainly oriented towards cultivation as the name implies itself. The net area sown possesses dominating to predominating role.

Acknowledgements

The author is grateful to Dr. (Mrs.) L. Shukla and Dr. H.S. Sharma, for providing guidance and encouragement throughout this study.

REFERENCES

Commonwealth Scientific and Industrial Research Organisation (1964): General Report on Lands of the West Kimberly Area, Western Australia, *Land Research Series No. 9*, Australia, p.10.

Davidson, D.A. (1980): *Topics in Applied Geography — Soils and Land Use Planning*, Longman, London.

Dent, D. and A. Young (1981): *Soil Survey and Land Evaluation*, George Allen and Unwin, London.

Doi, R.D. (1987): Land System, Landuse and Land capability — A case study of the Morel river basin (Rajasthan), Unpublished Ph.D. Thesis, University of Rajasthan, Jaipur.

Jainendra Kumar (1985): *Landuse Analysis*, Inter-India Publications, New Delhi.

Kostrowhicki, J. (1976): "On the methods of determining landuse, crop and livestock combinations as used in the polish studies", In V.C. Mishra, N.P. Ayyar and Pramila Kumar (ed.) *Essays in Applied Geography, in memory of the Late Professor S. Muzafar Ali*, University of Saugar, Madhya Pradesh, pp.111-20.

Mitchell C. (1977): *Terrain Evaluation*, Longman, London.

Mohmmad. N. (ed.) (1981): *Perspective in Agricultural Geography*, Concept Publishing Company, New Delhi, Volume One to Five.

Sharma, H.S. (1987): *Tropical Geomorphology — Morphogentic Study of Rajasthan*, Concept Publishing Company, New Delhi.

Siderus, W. (ed.) (1986): *Land Evaluation for landuse Planning and conservation in slopping areas*, International Institute for land Reclamation and Improvement, Wegeningen, the Netherlands, Publication 40.

Singh, A.N. (1984): "Delineating ravinous areas on aerial photographs and satellite imagery — A comparative study, *Photonirvachak Dehra Dun*, Vol. 12 (1), pp. 49-54.

7

Problems and Potentials of Pedogeomorphic Surfaces in the Basin of Purna

R.M. POFALI, J.L. SEHGAL and M. SWAMINATHAN

Introduction

INVENTORY OF the natural resources is prerequisite for judicious management of the basin which is a natural geographical unit. The land, soil and water are the important resources of Basin. Scientific management is imperative for optimum utilisation of these resources to the best of their capabilities without disturbing the ecological balance. Basin planning requires comprehensive understanding of the environmental characteristics and pedogeomorphic surfaces which are the products of the same geomorphic processes. The assessment of problems and potentials of each geomorphic surface is vital for developing a better land use planning befitting its environmental and pedogeomorphic characteristics.

The objectives of the present study are to identify and characterise the different pedogeomorphic surfaces of Purna Basin, assess their problems and potentials and suggest measures for their better use without degradation effects.

Methodology

With the above objectives in view, a geomorphic analysis map for the basin of Purna was generated using false colour composite on scale 1 : 250,000 supplemented with topographical maps. Subsequently field traverses were undertaken to study characteristics of each geomorphic surface to establish soil landscape relationship. Problems and potentials were assessed based on pedogeomorphic characteristics and physical environment of the basin. Critical evaluation of these features and soil analysis data led to suggestions on land utilisation pattern under judicious management practices.

Results and Discussions

The Purna basin, one of the important cotton growing regions of India, is located between 76°0′ to 77°55′E longitude and 20°10′ to 21°41′N latitude covering an area of about 18300 km² (Fig. 7.1). The Purna is a tributary of the westerly flowing Tapti river; the major portion of Purna basin falls in Maharashtra. This faulted basin is bounded by the Satpura scarp in the north and Ajanta scarp in the south.

The basin enjoys tropical monsoon climate with mean annual rainfall ranging from 800 mm in the south to 1500 mm in the north. The mean annual temperature ranges from 22°C to 27°C. The basin experiences moderate summer and cool winter in the north and south while the central plain experiences opperassive summer and mild winter.

The geological succession consists of basalt of the Tertiary period and alluvium of sub-recent to recent Quarternary period. The elevation of the basin ranges from 250 to 900 metres above the mean sea level. The general inclination of the terrain is from west to east with subsequent slopes from North to South and *vice versa*. Landscape analysis carried out from the false colour composite and supplemented with field observations reveal that the basin has three well-marked geomorphic surfaces which differ from one another in respect of morphology, morphometry, pedogenesis and landuse patterns. They are (1) Northern hilly terrain with shallow clayey soils. (2) Central alluvial plain with deep to very deep black clayey and alluvial soils and (3) Southern plateau surface with very shallow to shallow clayey soils. Pedogeomorphic characteristics of these surfaces are discussed; their problems and potentials highlighted and ameliorative measures suggested for optimum utilisation (Fig. 7.2).

1. *The Northern Hilly Terrain with Shallow Clayey Soils:*

This geomorphic surface is located on the northern fringe of the basin. The elevation ranges from 450 to 900 metres above the m.s.l. This unit stands out prominently on the physical landscape

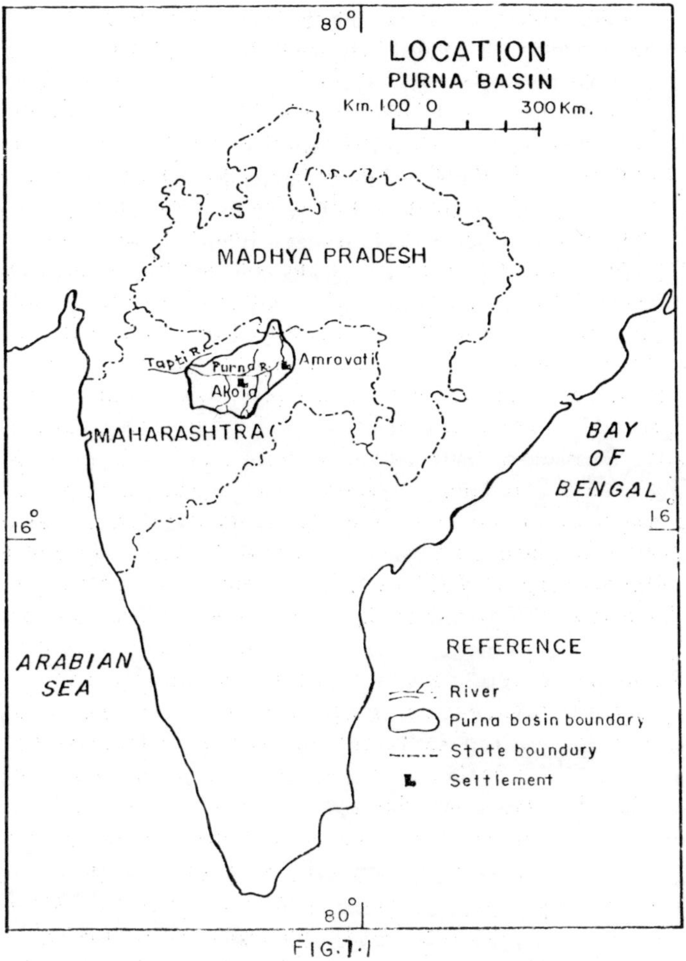

FIG. 7.1: Location of Purna Basin.

of the basin. The bold ridges and hills are composed of basaltic flows with trappean characteristics. The surface of this unit exhibits steep to very steep slopes and abrupt fall towards the Central part. Under the present climatic conditions, trap weathers with characteristics spheroidal exfoliation giving rise to rounded boulders of various sizes scattered on the outcrops. Prismatic jointing and columnar structure are observed in basalt along the hill side as well as on sloping surfaces. Severe dissection of the hill ranges have created narrow and deep valleys separated by escarpment slopes. The scarp slope facing towards south is severely dissected by the radiating streams. The terrain falls under high rainfall zone ranging from 1200 to 1500 mm. The mean daily maximum and minimum temperatures in the month of December are 22°C and 13°C and in May 35°C and 23°C respectively, indicating cool winter and mild summer. The terrain is clothed with tropical dry deciduous forest with teak as one of the most important commercial species.

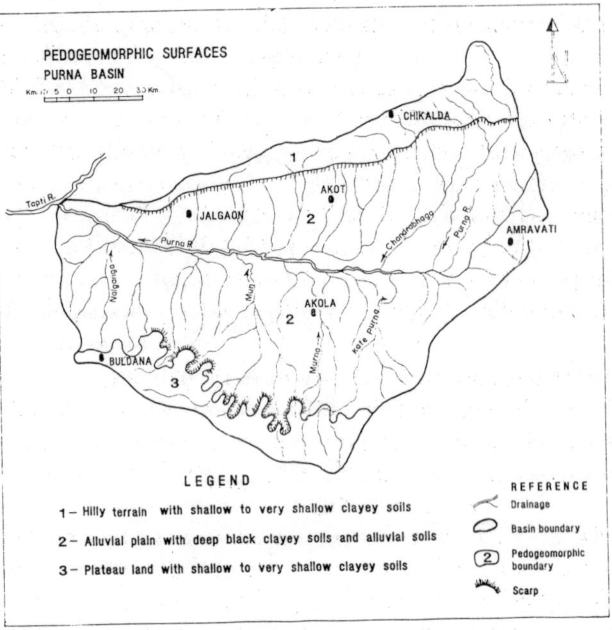

Fig.7.2

FIG. 7.2: Pedogeomorphic Surfaces of Purna Basin.

The rainfall and topography have significantly influenced soil formation. The soils are very shallow, dark brown to dark reddish brown, clayey skeletal, non-calcareous well to excessively drained, low in moisture retentive capacity and underlain by weathered basalt. They are acidic in reaction, rich in organic carbon and subject to severe erosion due to excessive topographic characteristics. They are less productive and unsuitable for agriculture (Fig. 7.3 & Pedon 1 Table 7.1).

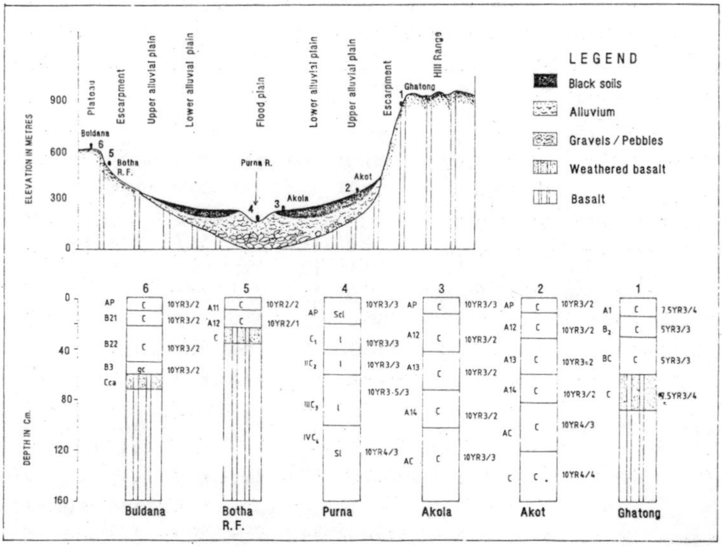

FIG. 7.3: Soilscape of Purna Basin.

Due to rugged characteristics of the terrain the surfaces are unstable and prone to erosion and excessive runoff. This has affected the pedological characteristics significantly. In spite of high rainfall the soils are droughty and the region is poor in soil and water resources. Low density forest cover the ranges facing the Purna valley. Deforestation in the recent past and shifting cultivation practised by the local tribals have aggravated the erosion. Most of the hill slopes are barren and grooved by monsoon gullies. The steep slopes, rapid runoff, shallow, gravelly and stony soils render agriculture almost impossible. The subsistence agriculture followed by some local tribals depends on cultivation of maize, minor millets and wheat.

TABLE 7.1: Physico Chemical Properties of Soils of Purna Basin

Location	Physico-graphy	Ped-on No.	Hori-zon	Depth (cm)	Particle size class and its range of diameter (mm)			Orga-nic car-(%)	Ca-Co_3 (%)	pH (1: 2.5) H_2O
					Sand (2.0-0.05)	Silt (0. (0.5-0.002)	Clay (0. 002)			
1	2	3	4	5	6	7	8	9	10	11
Ghatang	1 Hilly Terrain	1	A1	0-16	14.3	34.5	51.2	1.44	—	5.8
			B2	16-30	9.8	35.5	54.7	1.29	—	5.8
			BC	30-60	18.5	29.3	52.2	0.81	—	6.0
			C	60-90	36.4	22.8	40.8	0.54	—	6.2
Akot	2a Upper alluvial plain	2	Ap	0-10	17.0	24.3	58.7	0.48	6.5	8.2
			A12	10-35	12.5	28.0	59.5	0.48	6.3	7.9
			A13	35-60	12.7	27.4	59.9	0.47	7.0	7.8
			A14	60-83	11.9	27.9	60.2	0.43	7.0	7.8
			AC	83-122	16.3	28.0	55.7	0.35	12.1	8.0
			C	122-135	35.4	20.1	44.5	0.16	15.8	8.0
Akola	2b Lower	3	Ap	0-17	12.5	33.0	54.5	0.61	5.7	8.0
			A12	17-46	12.6	31.2	56.2	0.40	5.8	8.1
			A13	46-73	10.1	30.7	59.2	0.37	6.0	8.0
			A14	73-113	14.9	25.4	59.7	0.35	6.0	8.0
			AC	113-142	14.2	25.9	59.9	0.29	7.9	8.1
Purna	2c Flood plain	4	Ap	0-20	19.25	35.44	37.84	0.63	6.16	7.7
			C1	20-44	36.80	31.36	25.60	0.42	6.07	7.9
			IIC2	44-66	62.10	20.40	13.20	0.08	6.95	8.2
			IIIC3	66-104	47.70	28.80	18.80	0.19	14.08	8.7
			IVC4	104-151	65.11	18.20	12.80	0.05	11.17	8.7
Botha R.F.	3 Plateau Scarp	5	A11	0-10	14.2	19.3	66.5	1.24	—	6.6
			A12	10-24	18.0	19.5	62.5	0.99	—	6.7
			C	24-38	Weathered basalt.					
Buldana	Plateau top	6	Ap	0-10	15.6	18.5	65.9	0.51	3.3	7.2
			B21	10-33	13.4	27.1	59.5	0.48	3.6	7.4
			B22	33-54	11.5	37.0	51.5	0.48	3.6	7.5
			B3	54-60	25.5	29.5	45.0	0.43	10.9	7.8
			Cca	60-72	Weathered basalt mixed with powdary lime					

On the basis of pedogeomorphic characteristics, soils key out to Land Capability Class VI to VIII and suitable for forestry, pasture and wild life preservation. The area exhibits favourable sites for reservoirs which may be utilised to collect runoff water for irrigating the lower areas. The area is endowed with picturesque landscape having mild summer characteristics. This may be developed as landscape for tourism and summer holiday resort with recreation facilities and encouraging wild life sanctuary. At places the terrain is potential for high altitude horticultural crops like tropical fruits, shrubs and flowers propagation of which is beneficial to the impoverished farmers.

Soil and water conservation to restrict runoff and erosion, water harvesting, afforestation and gully plugging are the important measures to be taken to maintain the ecological balance in the region.

2. *Alluvial Plain with Deep Black Soils and Alluvial Soils*

This gently sloping to flat plain enclosed by two fault scarps of the northern hills and southern plateau is a geomorphic surface of deposition. It occupies 14750 km² which is 3/4th area of the basin. The elevation of this unit ranges from 250 to 400 m above m.s.l. It tapers in its width from 60 km in the east to 10 km in the west. The area receives rainfall of 800 to 1000 mm and forms a transitional zone between humid region of the north and semi-arid region of the south. The mean daily maximum and minimum temperatures in December are 29.3°C and 11.9°C and in May 42.4°C and 27.5°C respectively. The region experiences hot summer and mild winter.

The numerous sub-parallel gullies and hill tarrents descending from the north and south escarpments deposit huge amount of sediments due to sudden loss of gradient. The streams on the plain have a shifting tendency and develop meandering courses, abandoned channels, scars and oxbow lakes. The alluvial plain is badly dissected due to intensive gully erosion giving rise to badland topography. The plain is an assemblage of alluvial fans, interfluves and flood plains. The plain may be further subdivided on the basis of local topographic forms, heterogeneity in soil characteristics and cropping pattern. The Soil of the basin may be grouped to capability class II, III and IV.

2a. *Upper Alluvial Plain (Piedmont zone)*: This zone lies parallel to the north and south scarp faces and is 10-15 km in width. The numerous monsoon gullies descending down the steep scarp slopes of the Satpura hill and Ajanta plateau have deposited alluvium at the foot of the scarp which on coalescence have formed a piedmont. Descending from the high altitude, these streams have changed to sluggish meandering streams and deposit the sediments due to loss of gradients.

Soils over the piedmont zone are medium deep to very deep, dark brown to very dark greyish brown at the surface grading to dark yellowish brown sub-soil horizon clayey, highly calcareous and well drained, with good moisture retentative capacity underlain by stratified reddish/yellowish material derived from debris washed down, from higher reaches indicating lithological discontinuity. They are alkaline in reaction and low in organic carbon. They are fertile and respond well to irrigation and have potential for intensive agriculture. (Fig. 7.3 & Pedon 2 Table 7.1).

The problem of this pediment zone is severe gully erosion, shifting of stream courses and flooding by torrential streams in the rainy season descending from the adjoining scarps.

Deep soils, crop response to management and abundant ground water recharge distinctly indicate of high agricultural potential of this region. The soils have favourable texture and due to porosity of subsoil they are free from waterlogging and salinity.

Good soil tilth favour root penetration and optimum air and moisture to the plants. The area is well suited for horticultural crops and is famous for multiple and double cropping. Oranges, banana, sugarcane, tobacco, grapes, chillies, vegetables in addition to cotton, sorghum and wheat are usually raised on well-irrigation. Betel vines is a subsidiary garden crop. This is one of the potential agriculture regions of the Purna basin which is green throughout the year.

Training of river courses and gully plugging will save this valuable agricultural land from erosion hazard and sediment deposits. Use of soil based agrotechnology, improved management practices, drip irrigation, etc., will boost both agricultural and horticultural production.

2b. *Lower Alluvial Plain (Interfluvial zone)*: Adjoining the flood plain lies a vest tract of gently sloping to flat alluvial surface at an elevation of about 300 m above m.s.l. This surface is drained by the numerous sub-parallel streams. It is covered by deep to very deep black soils and is one of the major cotton growing tracts of the Purna basin. Water-table is very deep and there is prevalence of salinity is sub-soil water. Water is generally brackish on account of leaching of salts and calcareous material into the deeper horizons. At places salt encrustations are seen on the surface. Slow permeability of sub-soils creates waterlogged conditions during west season.

The soils of this zone are deep to very deep, very dark greyish brown to dark brown, calcareous, moderately well to ill-drained, high in moisture retentative capacity, and high in shrink swell potential with smectite as the dominant clay mineral. They develop slicken sides in the sub-soil layers. These soils are alkaline in reaction and alkalinity increases with depth. The soils are fertile in plant nutrients such as lime and magnesium and have high production potential. They are low in organic matter, nitrogen and phosphorous. These soils exhibit deep and wide cracks in summer giving rise to gilgai micro-relief on the surface. The average clay contents ranges from 41 to 51 per cent (Pedon 3 Table 7.1) and shows increasing trend with depth. The effective rooting depth varies from 70 to 85 cm. The soils below solum remain moist while at the upper part moisture remains at wilting point during hot dry period (Fig. 7.3 & Table 7.1).

The major problems of this geomorphic surface are related to erratic rainfall, flat relief, slow permeability, waterlogging, salinity and brackish groundwater.

During high rainfall years the crops suffer from waterlogging and later from salinity due to rise in groundwater table. The flat relief condition creates problems of drainage and water remains stagnant on the surface for a long time making agricultural operations difficult. In low rainfall years or on drying the soils become dry and hard and develop deep and wide cracks resulting into loss of stored moisture during maturity stage of the crop. In a year of well-distributed rainfall the yields of crops are encouraging. Imperfect drainage, poor tilth, high bulk density, slow intake rate and salinity hazards are the major constraints of these soils. Deep and brackish groundwater and slow permeability render the otherwise fertile soil difficult for irrigation.

In normal years of good rainfall distribution these soils have potential for production of cotton, sorghum, pulses in *Kharif* and wheat in *Rabi* on stored moisture. Under irrigation by canal these soils are very potential for sugarcane, citrus, spices, and vegetables and other fruit crops provided drainage requirements are met.

The management of these soils is very important for good harvest. Regular ploughing, application of green manure, provision of drainage, introduction of fresh water canal irrigation, judicious irrigation, soil and moisture conservation, gully plugging, bunding, introduction of salt tolerant crops, relay cropping are some of the ameliorative measures to manage these soils and to save them from further deterioration.

2c. *Flood Plain*: This zone is located along the banks of the Purna and its major tributaries. It lies at an elevation of 250 to 300 m above m.s.l. The zone is composed of flood time deposits, derived from encircling high lands. The alluvium is fairly thick (about 40 m deep) highly calcareous and mixed with boulders and pebbles of basaltic country rocks/and *kankar* together with some rounded and earthy material. Much of this alluvium produces efflorescence of salts of soda chiefly and well sunk in the vicinity yields brackish water. This unit is highly dissected by numerous gullies giving rise to badland topography.

The soils associated with this subdivision are brown to dark brown, very deep, well-drained, moderately permeable, lighter in texture (fine loamy) and highly calcareous with deep sub-soil of sandy loam texture. The soils show irregular textural pattern with packets of gravel and lime. The soil profile is stratified; clay content is irregularly distributed throughout the profile and similar is the distribution of organic carbon and coarse fragments. They are mildly to moderately alkaline (Fig. 7.3 & Pedon 4 Table 7.1).

Major problems of this subdivision are seasonal flooding, shifting of river courses, severe bank cutting and brackish water. The area hinders use of machinery due to dissected and hummocky topography. With proper measures the area is potential for raising vegetables and garden crops in addition to sorghum. Banana, guava, *ber* etc. are profitably grown in this subdivision. The Soil characteristics are favourable for irrigation provided water supply is made available through left irrigation from the adjoining river and stream beds constructing check demas. Underground water being brackish cannot be used for irrigation in this tract. Under dry land farming, *Kharif* sorghum followed by wheat adopting suitable soil and moisture conservation measures will ensure higher profitability. Relay cropping introducing vegetables, oilseeds, and fodder crops are suggested besides groundnut and pulses. Under irrigation these soils are potential for raising three crops, i.e., *Kharif* sorghum, *Rabi* wheat followed by vegetables. These soils are best suited for vegetables along the river courses. In summer in the beds of major rivers raising melon and *tinda* is profitable. Sugarcane, fruit crops like oranges, papaya, banana etc. can be introduced under irrigation forming.

In severely gullied area levelling, gully plugging, training of river courses, development of grassland and plantation of agave to check surface erosion and adoption of green manuring for stabilising soils and maintaining fertility are imperfective.

3. *Plateau Surface with Shallow Clayey Soils*

Southern boundary of the basin is marked by the plateau of Ajanta with an elevation of 450 to 600 m above m.s.l. and has given rise to undulating to rolling terrain. The plateau presents an escarpment on the its northern slope facing the Purna basin. This escarpment is less well-defined than the Satpura scarp and form a rough hilly country with steep gradients and is deeply scoured by a number of gullies and streams that become torrential during the rains and active recession of the fault scarp due to sub-aerial denudation. The plateau is composed of horizontal sheets of basaltic flows which are well-exposed in the scarp section and subject to spheroidal weathering.

Climatically this geomorphic surface falls in tropical semi-arid monsoon type with well expressed rainy, winter and summer seasons. The average rainfall of the plateau section is 800 mm. Mean daily maximum and minimum temperatures are 38.3°C and 26.2°C in the month of May and 27.6°C and 15.1°C in the month of December respectively. This indicates prevalence of moderate summer as compared to oppressive heat of the plains. The plateau falls under low rainfall zone and is covered by tropical dry deciduous and tropical thorn forests; teak is an import commercial species.

Soils on the scarps are skeletal, very shallow, stony, very dark brown, clayey, noncalcareous and underlain by weathered basalt. The soils are neutral to alkaline and are rich in organic carbon as compared to the soils of alluvial plain and subject to severe erosion (Fig. 7.3 and Pedon 5 Table 7.1).

The problems of this pedogeomorphic surface is steep gradient, rolling character of the terrain, low rainfall, excessive runoff, erosion, droughtiness, deforestation and shallow rooting depth of the soils. As a result these surfaces are not suitable for agriculture and fall in capability class VI and VII. They can be used for forestry and permanent pasture. Harvesting surface water by putting checkdams in the narrow valley section of the escarpments is a distinct possibility.

Soils on the gently sloping flat tops of the plateau are shallow, clayey, very dark greyish brown to dark brown underlain by calcareous weathered basalt. They are alkaline in reaction. They are low in nitrogen and available potash and have moderately good fertility status and good moisture holding capacity (Pedon 6 Table 1).

The problems of this pedogeomorphic surface are low rainfall, rugged terrain characteristics, shallow rooting depth of the soils, high runoff, erosion, droughty condition, low groundwater and lack of irrigation facilities.

The plateau surfaces is suitable for Rabi cropping. Safflower is important among the oilseeds in addition to pulses.

Soil and moisture conservation, gully plugging, contour bunding, and cultivation of shallow rooted crops are important measures to increase biomass production.

Conclusion

The study of pedogeomorphic surfaces of the Purna basin indicates that the northern hilly region is potential for development of forestry, grasslands and sanctuary. It has very low potential for agriculture due to shallow soil cover and steep gradient inducing high runoff and erosion. There are numerous picturesque sites which may be developed as picnic spots and the narrow valley bottoms may dammed for storing surface water. Cool climate in the summer may attract tourists as is the experience at the hill station Chikaldara. However, more development is necessary for boosting tourism.

The alluvial plain is one of the most fertile regions of the basin and rich in soil and water resources. The stored water in the hilly and Plateau sections may be utilised for boosting agriculture in the plains. In the foot hill zones the fresh groundwater is abundant and soils are potential for developing double and multiple cropping system. New technology and agronomic practices will help yield profitable harvest in this tract. In the deep black soil zone the management of soil and water resources are important, as the groundwater is brackish. Judicious Canal irrigation may boost agricultural production. The flood plain surfaces are prone to erosion and flooding during rainy season. The soils are suitable for development of vegetable and garden crops if they are managed properly. Gully plugging, sedimentation and training of stream courses are imperative in this unit to save the fertile land from further degradation.

The scarp of the plateau surfaces should be protected by restoring vegetal cover and the narrow valley bottoms may be used for harvesting runoff water. Afforestation, contour bunding, development of grasslands and gully plugging are some of the important measures to be adopted to save the plateau surface from devasting erosion hazard. The climatic characteristics of the plateau are favourable for Rabi crops. Soil and moisture conservation measures are important due to low and erratic rainfall characteristics. Efforts may be made to bring irrigation water on the plateau surface for intensive agriculture.

Acknowledgement

Authors are grateful to Dr. L.R. Hirekerur, Sr. Scientist, NBSS & LUP, for his help in preparation of this paper. Thanks are also due to Dr. K.S. Gajbhiye, Sr. Scientist, Dr. R. Srivastava, Scientist and Sh. N.Y. Parhad, T-4 for their help in the field and laboratory.

REFERENCES

Krishnan M.S. 1960: *Geology of India and Burma*, Higginbothams Pvt. Ltd., Madras.

Maharashtra State Gazetteer of India 1951 : Amravati, Akola, and Jalgaon districts.

NBSS & LUP 1980: *Soils and Suggested Land Use of Maharashtra*, Bull. No. 4, ICAR, National Bureau of Soil Survey and Land Use Planning, Nagpur.

8

Flood Hazards and River Bank Erosion in the Lower Damodar Basin

PROBHAT KUMAR SEN

THE PROBLEMS of flood and bank-side erosion were typical chronic problems in the flood plains of the lower Damodar basin in West Bengal. The present paper deals with the flood hazards and river-bank erosion of the lower Damodar, the longest tributary from the Chota Nagpur plateau flowing to the Bhagirathi. The river Damodar (541 km) originates from the Khamarpat hill (1,366 m) in the Chota Nagpur plateau of Bihar and flows in an easterly and southeasterly directions through the coal-bearing Gondwana beds of Bihar and West Bengal and below Panagarh it merges into the alluvial plains with a conspicuous change of its course to the south at Kamalapur village near Burdwan. Further below Jamalpur it bifurcates into two channels, Mundeswari and the main Damodar channel known as Amta Channel with a close third distributary, the Kana Damodar with its silted-up off-take, but connected at present with the help of canalisation. At present the Mundeswari channel, carries 90 per cent of the monsoon discharge whereas, the main Damodar, (Amta channel) carries only the remaining 10 per cent. The former meets the river Rupnarayan south of Amta while the latter the river Bhagirathi at Falta. In the northern part of the lower catchment, immediately below the apex of the Damodar deltaic plains near Panagarh, a number of distributary channels such as Khari, Banka, Behula, Gangur etc. in the north and Ghea, Kunti, Ilsura etc. further south flow towards the east to the river Bhagirathi having lost their connection with the mother channel, the river Damodar. Most of these channels such as Khari, Banka, Behula and Gangur are considered as the past channels of the main Damodar during the 17th and 18th centuries and thereafter during the late 18th and 19th centuries the main flow was through the Ghea, Kunti, Kana and Kana Damodar and at present through Amta channel and river Mundeswari.

The whole system of drainage of the river Damodar extends over an area of 37,000 sq km of which the upper catchment in Bihar constitutes 63.6 per cent area whereas lower catchment in West Bengal, only 36.4 per cent. The wide elliptical upper catchment tapers down to a narrow elongated basin in the form of a funnel which paved the path of heavy onrush of surface runoff of the upper catchment to the lower with the obvious chronic hazards of annual flood and valley-side erosion in the lower catchment during the pre-dam period before the implementation of the first phase of the multi-purpose Damodar Valley Project. The age-long notoriety of floods in the lower Damodar causing immense damage to the agricultural economy and settlements had earned the river Damodar the disrepute of being known as the 'river of sorrow'. In the newly emerging scene, the flood hazards and river bank erosion of the lower Damodar have been considerably mitigated after the construction of the four dams and reservoirs and later Tenughat Dam on main Damodar river in the upper catchment facilitating controlled release of water downstream. But floods still now pose serious problems. In this study the northern limit of the lower Damodar basin is taken from the confluence of the Damodar with its longest tributary, the Barakar near Damodar bridge site in Burdwan district immediately below the Maithon and Panchet dams on the Barakar and Damodar rivers respectively.

Channel Pattern and Hydrological Characteristics

The lower Damodar basin inclusive of the catchments of the northern beheaded distributaries extends over an area of more than 10,000 sq km and depicts the hydrographic characteristics of a moribund delta manifested in the nature of its branching system radiating out from the proximity of an apex which have undergone considerable oscillations and changes. Unlike the river Bhagirathi and its distributary channels, the lower Damodar system of channels are less sinuous in nature maintaining remarkably straight courses. Lithologs in the lower basin testify the profuse sedimentation of Sandy materials brought down from the Chota Nagpur highlands and the presence of the abandoned channel courses. The whole length of the main Damodar channel below the dams is only a marginally meandering channel (1.46), with remarkably straight course up to Burdwan (1.04) and only wandering

(1.12) in the lower reach. The other lower distributary channels, e.g. the Mundeswari (1.24), Amta channel (1.10) and Kana Damodar (1.18) are also characterised by the straightness of their courses with S.I. ranging from 1.0 to 1.25 whereas the northern distributary channels are definitely meandering channels with values of 2.0 and above with the exception of Banka (1.37). The senility, finer grain size and flow over the gentler slopes of the younger Deltaic plains obviously initiated the incipient growth of wandering which degenerated into meandering channels in the older stage.

The total length of the lower Damodar from the confluence of Damodar with Barakar is 250.15 km. The distributaries vary in lengths from 210.07 km in case of Khari, 100.79 km of Banka the northernmost channel to extremely short lengths of some channels like Faringachi 21.53 km and Gurjali only 9.78 km.

While the elevation at source of the lower Damodar at the northern end near the Damodar Bridge site is only 87.92 m, in case of certain channels it is less than 6.09 m. Slope is also very gentle either less or about 1′ only. The observed long profiles of bed-levels of the channels typify aggradational characteristics which are conducive to spilling of the channels and widespread innundation of the flood plains. As regards the water conveyance characteristics, the W/D or F ratio is one of the simplest but most powerful index which unravels the relative importance of vertical and lateral erosion. The values of the lower Damodar system of channels except for some northern distributary channels are abnormally high indicating features of shallow wide channels fraught with the processes of lateral erosion and aggradation. In case of lower Damodar, these high values range from 56 to 583, arising out of low gradient, non-cohesive sandy bed and monsoon flushing phenomenon. Some sections of river Mundeswari and Amta channel also possess such high values ranging from 100 to 366 in case of the former whereas 70 to 410 in case of the latter. The other distributary channels, especially in the northern section, have considerably lower values ranging from about 7 to 25 owing to silty channels along with tectonic movement. After Crickmay, the ratios in between 14 and 28, with a common value of 20 are most representative features of a channel in natural alluvium but in case of 40 or above, the channels are likely to develop impotency. Thus most of the lower Damodar channels have poor drainage capacity and any high discharge cannot be disposed of easily with inevitability of floods and bank-side erosion.

History of Floods and Modern Achievements

The hazards of floods in the lower Damodar basin have been almost an annual occurrence in consequence of high discharge conditions and cyclonic rainfall or rainstorms during the monsoon in the upper catchment and low topographic depressions with inadequate indigenous devices to counteract them. Thus it has been a common practice to survive with floods till the implementation of the modern integrated Damodar Valley Project. Any discharge above 250,000 cusecs (7075.0 cumecs) at Rhondia (the only station with the longest record over a century) was fraught with floods.

E.L. Glass's Report on 1918 submitted to the Government of Bengal assessing the flood condition during the 61-year period from 1857-1917 at Ranigunj estimated the occurrence of extremely abnormal high flood above 450,000 cusecs (12735 cumecs) only once, whereas abnormal floods with a peak discharge above 300,000 cusecs (8490.0 cumecs) as 20 per cent of the total occurrences, those between 200,000 to 300,000 cusecs (5.66 to 8.49 thousand cumecs) 55 per cent and those below 200,000 cusecs, the subnormal floods, only 25 per cent of the total.

With a mean peak discharge of 275,000 cusecs (7782.5 m^3/sec) at Rhondia during 1933-57, before the construction of the Maithon and Panchet dams and reservoirs in 1957 and 1959 respectively, the Return Period of 100,000 cusecs was a 1.36 year flood, 300,000 cusecs a 3.37 year flood, 625,000 cusecs a 13-year flood, and 50,000 cusecs a 26-year flood. In spite of these Return Periods, occurrences did not obviously follow any orderly intervals.

The incidences of floods of varying magnitude and areal extent were invariably in and around Burdwan town on the river Damodar and in the further downstream section. The flood plains of the distributary channels also experience widespread occurrence of floods besides the trans-Damodar section with its saucer-shaped low topography, high sub-surface water yield, occasional rainstorms and high rainfall with the high discharge from the up-stream section. Unfortunately no satisfactory studies have been conducted partially due to the absence of recording stations of long standing.

The flood hazards had been encountered traditionally by raising embankments both along the left and right banks of the river Damodar and its lower distributary channels. The right bank embankment had to be demolished sometimes to mitigate the floods in the built-up areas on the left bank. The embankments are being maintained even after the construction of dams and reservoirs and extended over the lower distributary channels.

The post-dam period with a flood protection benefit of 1.041 million acre feet falls short of the original recommendation of Mr. W.L. Vorduin of 2.90 million acre feet and the planned target of 1.510 m ac ft. in the first phase of the construction. The Maithon and Panchet dams and reservoirs are of immediate concern to the lower catchment which provide combinedly the flood moderation capacity of 852,000 ac ft against a recommended capacity of 1,321,000 ac ft thereby providing the flood benefit of only 6.40" (162.56 mm) of surface runoff in place of 18" (45.20 mm) as advocated by Mr. Vorduin. The peak discharge condition during the post-dam period (1959-75 = 19 years) shows the incidence of two high floods above 300,000 cusecs at Rhondia in 1959 (3.1 lakh cusecs) and in 1978 (3.85 lakh cusecs). The R.I. of the floods of different magnitudes during these 17-year period (1954-75) shows 1.16 lakh cusecs as a 2-year flood, 2 lakh cusecs as a 9-year flood and 3.10 lakh cusecs as a 18-year flood.

The application of V.T. Chow's formula regarding the R.I. of floods during the pre- and post-dam conditions shows the diminishing trend conspicuously and that of Gumbel's law of maximum value, though of no significance in case of controlled release of water from the dams, depicts a similar situation.

The hazardous floods of October 1959 and September-October 1978 obviously prove the inadequacy of the flood protection benefits during the post-dam period. Consequent upon a release of 288,000 cusecs (8150.4 cumecs) from the dams and with the uncontrolled surface runoff up to Durgapur, the peak discharge attained a value of 350,000 cusecs (9800 cumecs) at Durgapur and caused widespread floods in the lower and trans-Damodar catchment causing immense damage to agriculture, settlement and lives. Following 1959, the 1973 peak discharge of 2.02 lakh cusecs at Rhondia, owing to the cyclonic rainfall, generated considerable flood havoc. The flood during September-October, 1978, however, created the worst flood havoc not only in the history of floods of lower Damodar catchment but also over large parts of West Bengal. The cyclonic rainstorm during the 27-29 September, 1978 followed by a secondary peak in 2-4 October, 1978 caused heavy precipitation above 500 mm both in upper and lower catchments which necessitated the release of 1.63 lakh cusecs of water combinedly from the Maithon and Panchet dams which further augmented to 3.30 lakh cusecs at Durgapur and further below at Rhondia to 3.85 lakh cusecs causing widespread floods. The flood was not only confined to the lower and trans-Damodar basin but also in the Durgapur and surrounding areas which hardly ever experienced much flood hazards even before the construction of the dams owing to excessive rainfall in this small section of the catchment of about 2500 sq km (965 sq miles) below the dams up to Durgapur with 90 per cent or more surface runoff from the small tributary catchments in this undulating terrain. The condition in the trans-Damodar section further worsened as the high discharge from the upstream section synchronised with high rainfall above 500 mm in that catchment contributing a surprisingly high amount of sub-surface runoff yield. The Damodar with its tributaries in the upper catchment experienced the all-time highest flood in the lowermost reservoir points having a peak inflow of 851,000 cusecs (24,107 cumecs). Though the D.V.C. authorities claim a splendid performance of controlled release of water from the dams, the inadequacy of the dams for such an abnormally high runoff conditions is established beyond doubt. It is paradoxical to note that during the non-monsoon period these channels are almost dry with very low discharge of water. Any peak discharge above 70,000 cusecs is likely to cause the floods in the Mundeswari and Amta channels, which are further aggravated by the tidal phenomena in the lower reaches.

The flood protection measure has been achieved through the raising of the embankments along riversides, known as 'jacketting', from Burdwan downwards along both the Mundeswari up to Baxi near its confluence with Rupnarayan and Amta channels. In spite of embankments the spilling and breach occur very frequently. The other recommendation is for the resectioning of the channels for efficient flow. A number of investigations have been carried out before and after the implementation of the Damodar Valley Project to focus on these problems with poor practically no hydrological records. One of the questions regarding these channels is whether they should act only as flood discharge carrying channels deprived of their capacity to drain their own flood plains and catchments which are the most useful natural function of any channel in its catchment. Thus the areas adjacent to the embankments in the trans-Damodar catchment are transforming into ill-drained waterlogged areas.

It is being realised that the heavy siltation of the reservoirs are gradually diminishing the reservoir capacity with gradual diminution of flood cushioning benefit. At the present rate of heavy siltation in the reservoirs, the life of reservoir is apprehended to be of only 50 years. The accelerated erosion originate from the processes of deforestation, urbanisation and industrialisation for which however, measures of conservation have been undertaken by the Damodar Valley Corporation.

River Bank Erosion

The river bank erosion posed another serious problem along the Damodar and its distributary channels. Even in the post-dam period, there are many vulnerable sites of bank-side erosion of considerable intensity. Some of the records, (though of doubtful nature) for a span of about a century unravels the vehemence of the bank-side erosion in the lower and trans-Damodar section. Showing, however, appreciable contrasts of the upper and lower reaches of the lower Damodar. The former flows through non-cohesive sandy and coarse-textured areas whereas the latter through the silty and cohesive materials.

To site a few examples, the retreat of valley side slopes at Jamalpur near the bifurcation point during 66 year (1881-1956) amount to 310′ and 100′ on the right and left banks respectively, indicating an average retreat of 4.4′ and 1.4′ per year. It is, however, likely that some incidents of the most significant erosion occurred when thresholds were exceeded and thus average values may not be of considerable significance. Further downstream, in the trans-Damodar section with trapezoidal forms of cross-sections of the channels, the C.S. of a few sites from Jamalpur to Duffer Chak show an increase in width from 320 m to 488 m and decrease in depth from 9.1 m to 5.48 m (Source : D.V.C. Lower Damodar Investigation Report, 1957). On superimposition of 1988 and 1956 cross-sections, the important generalisations of bank-side erosion are as follows:

1. The shapes of the cross-sections of the channels in 1881 and 1956 maintained trapezoidal forms with an almost horizontal or slightly sloping bed.
2. Apart from the thalweg, the flood plain is slightly inclined creating large areas vulnerable to floods in consequence of high discharge condition.
3. Valley-side accretion of silt and erosion of river banks are predominant phenomena.

Further, downstream at Jerah Gauge (M.P. $51^1/_2$) (M.P. = Mile Post), Champadanga (M.P. $57^3/_4$), Duffer Chak (M.P. 63 $^1/_4$) and Amta (M.P. $79^1/_2$), the cross-sections of the valleys are also characterised by pronounced erosional and accretional features leading to either recession or progression of the valley sides. At Jerah Gauge and Champadanga, the progression presumably by the processes of lateral accretion of silt is of the extent of 450′ (137 m) and 400′ (122 m) i.e., an average of 6.6′ (2.0 m) and 5.8′ (1.76 m) respectively per year whereas retreat of slopes of the extent of 700′ (213.36 m) and 200′ (60.96 m) i.e., 10.0′ (3.04 m) and 2.9′ (0.88 m) respectively per annum. At Duffer Chak, right bank gave way to recession of about 4.4′ (1.34 m) and accretion of about 13.2′ (4.0 m) per annum. These average figure may be quite erroneous for the fact that most of these changes must have been associated with high discharge conditions and even the observational data may not be strictly of same sites. Nevertheless, these approximate observations unravel the conspicuous dynamism of the cross-sectional changes.

The processes of erosion and sedimentation have altered radically in response to the regulated streamflow conditions in the lower Damodar and its distributary channels. Immediately below both the Maithon and Panchet dams, the scouring action is conspicuous. During 1961-65, the annual valley fill and scour have been estimated to be only 2/3′ with only localised bank-side erosion on Barakar and on Damodar near Ranigunj.

Further downstream the lower Damodar distributary channels have embankments on both the sides and thus the bank-side erosion is confined to the river valleys at some vulnerable points causing at times breach of the embankments. But the flood protection benefits have considerably mitigated the problem and changes are only episodal.

Conclusion

Pre-dam Condition

The ravages of the floods and bank-side erosion in the pre-dam years were rather endemic in nature and the most devastating ones with peak discharges of 600,000 cusecs or more at Rhondia had occurred in 1823, 1840, 1913, 1935, 1941 and that of 1943 which initiated the idea of the multi-purpose Damodar Valley Project as in T.V.A. (U.S.A.). The flood discharges accompanied by breaches on both the left and right embankments caused the most serious damages to agricultural products, settlements, communication and loss of lives. Though no accurate quantitative estimates of the nature of loss are available, the approximate estimate of the damages caused by minor annual floods during pre-dam period amounted to about Rs. 2,550,000 of paddy crops alone in addition to another Rs. 500,000 for the damages caused to communication, building etc., thus amounting to a total loss of Rs. 3,050,000. It is obvious that the damages in terms of the present price index will be many crores of rupees.

Obviously the indigenous devices for counteracting these floods were extremely inadequate and the settlements grew up generally at a considerable distance from the river sides in order to avoid these unforeseen ravages.

Post-dam Condition

As already discussed, the lower Damodar basin has not outlived the hazards of flood. The high floods of 1959 and 1978 obviously testify the inadequacy of full flood protection in the upper catchment. The 1978 flood has revealed that in spite of the controlled release of water from the Maithon and Panchet dams, the uncontrolled runoff in the catchment below the dams may augment the discharge at Durgapur and Rhondia by more than one lakh (100,000) cusecs in case of heavy rainfall. In the trans-Damodar distributary channels also, the sub-surface water yield is surprisingly high in case of high rainfall as during the cyclonic rains of 1978 flood.

The nature of damage and socio-economic impact can be gauged from the September-October 1978 flood which is the most hazardous in terms of inundated area, damages to agriculture, settlements, communication, loss of human lives and livestock with complete dislocation of the normal life. The number of breaches of the embankment alone was surprisingly high in the catchment districts of Bankura-437, Burdwan-751 and Hooghly-182. While the right bank of Damodar was breached at a number of places inundating a large number of villages, the left bank embankment was not breached saving the towns of Burdwan, Memari, Jamalpur, Galsi, Bud Bud etc. which belonged to the chronically affected zone in the pre-dam days.

Though exact estimates of loss are not available, some of the assessments of the District Magistrates of the different districts portray the huge magnitude of the damages. The nature of damages to embankments, roads, bridges etc. amount to several crore of rupees and the total post-flood recovery relief in Burdwan district alone had been earmarked as Rs. 150 crores while in Howrah district damages of Rs. 142 crores have been estimated with maximum damage to houses followed by standing crops and roads.

In spite of the gainful consequences of soil conservation, flood mitigation, irrigation in the upper and lower catchments, it is evident that the flood problems of the lower Damodar and drainage problems of the trans-Damodar basin have not been completely ameliorated. Furthermore, the problems of the trans-Damodar have not been paid adequate heed from the very inception of the D.V.C., though a number of investigations and recommendations have been made in the post-dam period.

REFERENCES

Crickmay, C.H. (1974): *The work of the river*. Macmillan Press Ltd., London.

D.V.C., (1978): *D.V.C. - Its flood control capacity, Past, Present and Future* (A cyclostyled Report).

Leopold, Luna B, Wolman, M. Gordon and Miller, John P. (1969): *Fluvial Processes in Geomorphology*. Eurasia Publishing House (Pvt.) Ltd., New Delhi-55. First Indian Reprint.

Sen, S. (1954): Drainage Study of the lower Damodar Valley. *Ind. Jour. Power & River Valley Development*, Calcutta.

Sen, P.K. (n.d.): The Genesis of floods in the lower Damodar catchment. In *The Concepts and Methods in Geography*, (ed.) Sen, P.K., Institute of Geography, Deptt. of Geography, Burdwan University, Burdwan-713104.

Sen, P.K. (1979): The recurrence of floods in the lower Damodar basin. *Ind. Jour. of Power & River Valley Development*, Calcutta, January.

U.N. Econ. Com. for Asia and Far East. Flood Control Series, No. 16, bangkok 1960. *A case study of the Damodar Valley Corporation and its projects*.

Ward, R. (1978): *Floods: A Geographical Perspective*, Macmillan Press Ltd.

FIGURE 1

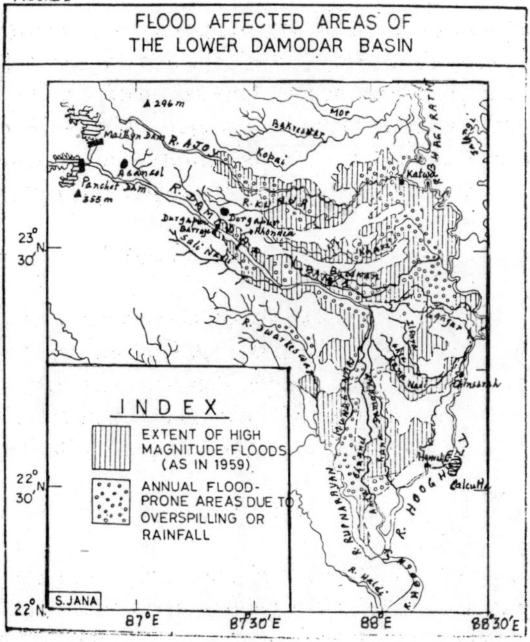

FIG. 8.1: Flood Affected Areas of the Lower Damodar Basin.

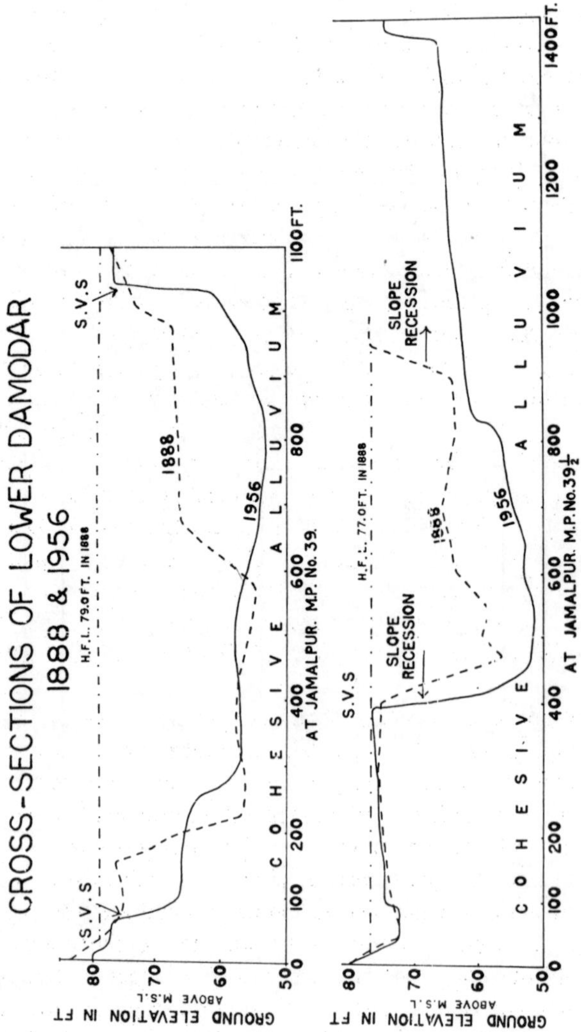

FIG. 8.2: Cross-sections of Lower Damodar, 1888 & 1956.

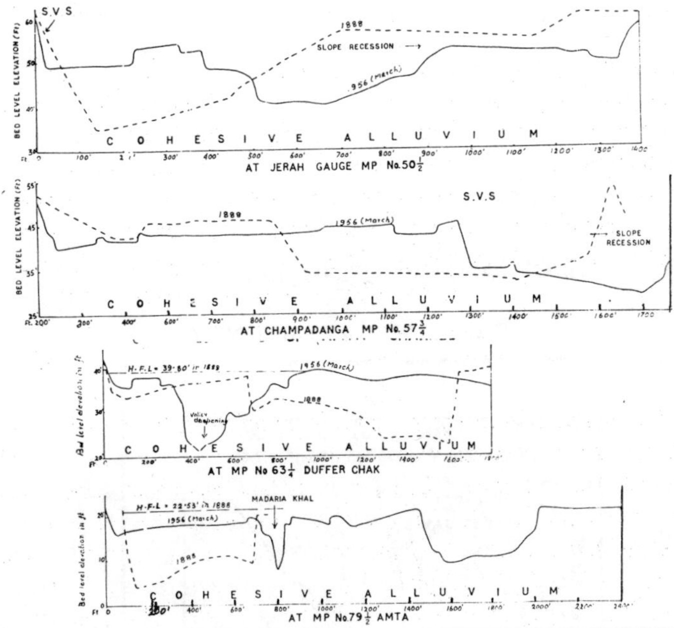

FIG. 8.3: Cross Sections of Amta Channel.

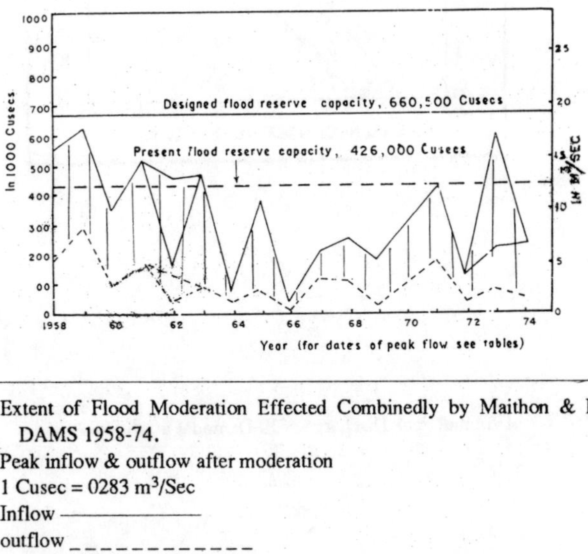

FIG. 8.4: Extent of Flood Moderation Effected Combinedly by Maithon & Panchet DAMS 1958-74.

Peak inflow & outflow after moderation

1 Cusec = 0283 m³/Sec

Inflow ————————

outflow _ _ _ _ _ _ _ _ _ _ _ _

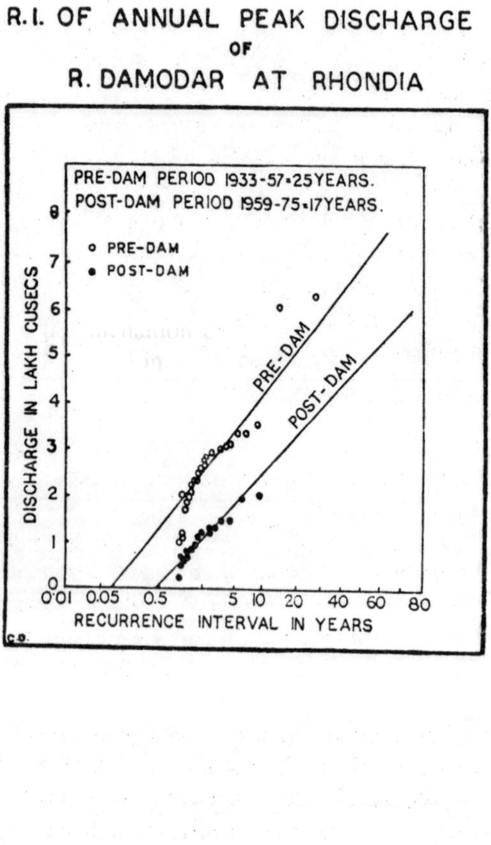

FIG. 8.5: R.I. of Annual Peak Discharge of R. Damodar at Rhondia.

9

Waterlogging – A Geo-environmental Hazard in Arid Land Irrigation : A Case Study of Ganganagar District (Rajasthan)

S.C DHIMAN and H.S. SHARMA

Introduction

CANAL IRRIGATION was in practice even during the civilisations of Tigris, Euphratus, Babylonia, Mesopotamia, Nile and Indus. The progressive mismanagement of land-water resources have contributed to the disintegration of past civilisations (Jacobson and Adam, 1958). Land-water resources if managed scientifically, irrigation leads to highly productive agro-eco-system. But if allowed to intrude upon the natural environment it violates the natural equilibrium of arid ecosystems and it can hasten the process of desertification including salinisation, waterlogging and other reversible/irreversible changes.

The Indus civilisation flourished in the valley of Vedic Saraswati River, presently Ghaggar river was extended over to this area (Tessitery, 1919; Stein, 1941; Herman, 1946 and Ghose, Kar and Zahid Husain, 1980). Desert gradually become inhospitable and remained no man's land for centuries (Man, Malhotra and Kalla, 1974). Superimposition of large scale surface water irrigation on this virgin arid land (Gang Canal, 1927; Bhakra Canal, 1954; and Indira Gandhi Canal, 1963) signified a complete anthropogenic reconstruction of the environment. Prior to the canal irrigation cultivation was confined either to dried river bed "Nali bed" with limited irrigation or topographic lows depending only on rainfall.

Geomorphologically, the area studied has been classified into aeoline and fluvial land systems. The characteristics and hazards associated with them are discussed in Table 9.1. Large inter-dunal flats in southern part are being utilised for storing excessive flood waters whereas northern part of the sandy plain is under rich agriculture.

Nalibed has been intensively developed for cultivation and is severely damaged with waterlogging hazards.

TABLE 9.1: Land Systems with Associated Environmental Hazards and Their Management

	Land form	Landuse	Hazards	Management
A.	**Aeolian Land System**			
i.	*Northern Sandy Plain*			
	Dunes low to med in height. No natural drainage, undulating topogrphy, range in height from 176 m to 206 m above m.s.l.	Intensive to poor	Wind erosion	GDC is to made impervious I.W.M. system is to be extend over entire area with intensive ground water development. Horizontal drainage is to be adopted along Indira Gandhi Canal.
ii.	*Southern Sandy Plain*			
	Dunes height to med dissected along the crest, and sheet common, large interdunal areas. No natural drainage, range in height from 172 m to 223 m above m.s.l.	Barren except topography lows, interdunal flats utilised as flood water reserviors.	Wind erosion	Flood link canals to be lined, impounding of flood water is not to be kept permanently and to be reduced, saddles to be made impervious plant species with higher water intake are to be grown.
B.	**Fluvial Land System**			
i.	*Ghaggar River Bed*			
	Flat, gently sloping towards west, High to medium dunes on the banks and occasional sand deposits over remaining areas, Meander scrolls, burred channels are poorly preserved.	Intensive to moderate cultivation with canal irrigation and and limited groundwater	Soil salinity and Waterlogging	Groundwater development to be taken up, I.W.M. system is to be extended over entire Ghaggar river bed.

(Contd.)

TABLE 9.1 – (*Contd.*)

Land form	Landuse	Hazards	Management
ii. *Drishadwati River Bed*			
Flat gently sloping towards Baropal from East and West, Meander scrolls are poorly preserved sand sheets common.	Intensive to moderate cultivation with canal irrigation.	Severe soil salinity waterlogging and marshy lands.	Surface water irrigation is to be minimised, canal are to be lined, Horizontal drainage is to be developed where water level is less than 1.5 m ground water from good quality shallow zone is to be developed, crops and plant species with higher water intake are to be grown.

I.W.M. Improved water management system is adopted in part of Indira Gandhi Canal command for controlled meticusily planned water distribution.
G.D.C. Ghaggar diversion channel.

Methodology

A multi-stage remote sensing approach is made with multi-date satellite imagery, aerial photographs (1968) and correlative ground truth data are interpreted visually for detailed studies for the impact of surface water irrigation on the land forms, surface runoff, groundwater regime along with the environmental changes for the period between 1968 and 1986.

Study area forms a part of the Thar desert and is located in southern part of Ganganagar district, Rajasthan covering about 2000 square kilometres. The area falls between 73° 50′ and 74°33′ east longitudes and 28°58′ and 29°20′ north latitudes.

Discussions and Results

With the superimposition of surface water irrigation, major part of flat and low topographic areas in both fluvial and aeoline landsystems have been transformed from waste lands to rich agricultural fields. Transformation of dunal uplands is still continuing. Large interdunal areas in the south east of Suratgarh are being utilised as natural reservoirs (Ghaggar Depressions) to store excessive Ghaggar flood waters. These reservoirs have been interconnected by cuts digging and saddles plugging of sand dunes.

With the filling of flood waters in the reservoirs over the highly permeable strata, sub-surface water regime was effected severely within a very short period. First the adverse effects like waterlogging, marshy lands and soil salinity appeared along the fluvial landforms along the old dried bed of Drishadwati river extending from Badopal (168 m above m.s.l.) towards Lakhawali in the east and towards Rangmahal in the west because of its lowest topographic location (altitude of the dried river bed ranged between 168 m to 170 m above m.s.l.) whereas the bed of depressions filled with water located at higher topography (altitude between 171 and 184 m above m.s.l.). Drishadwati river bed has started experiencing adverse conditions after the commission of Bhakra Canal (1954) and thereafter has further accelerated with introduction of Indira Gandhi Canal (1963). The increased intensity of Ghaggar floods since 1960 has also aggregated the adverse conditions. About 987 hectares was either under waterlogging or saline soils till 1968 and increased to 2800 hectares in 1972 along with the appearance of saline soils and marshy lands. These hazards have increased at an alarming rate after the filling flood waters depressions and by 1983, entire bed with a limited area under intensive agriculture has been transformed into a saline desert with ever waterlogging and marshy land condition. After damaging entire area in the north of depression, these adverse conditions are advancing towards west and north-west and are further aggregated from the excessive seepages from Indira Gandhi Canal. This has resulted in rise of water level of regional groundwater body and rest is more presently towards West.

Along the Ghaggar river bed, waterlogging condition has also been appeared between Talwara and Hanumangarh, around Pillibangam and Suratgarh. These are advancing further down stream. The rising water level around Suratgarh has started affecting the Suratgarh town.

In the aeoline lands system the waterlogging conditions has started appearing since 1975 in topographic lows along the Indira Gandhi Canal around Masitanwali and Shelwala Khurd villages and is extending towards West in the northern sandy plain in all direction from Ghaggar depressions. Since 1970 increment of waterlogging conditions in various periods had been given in Table 9.2.

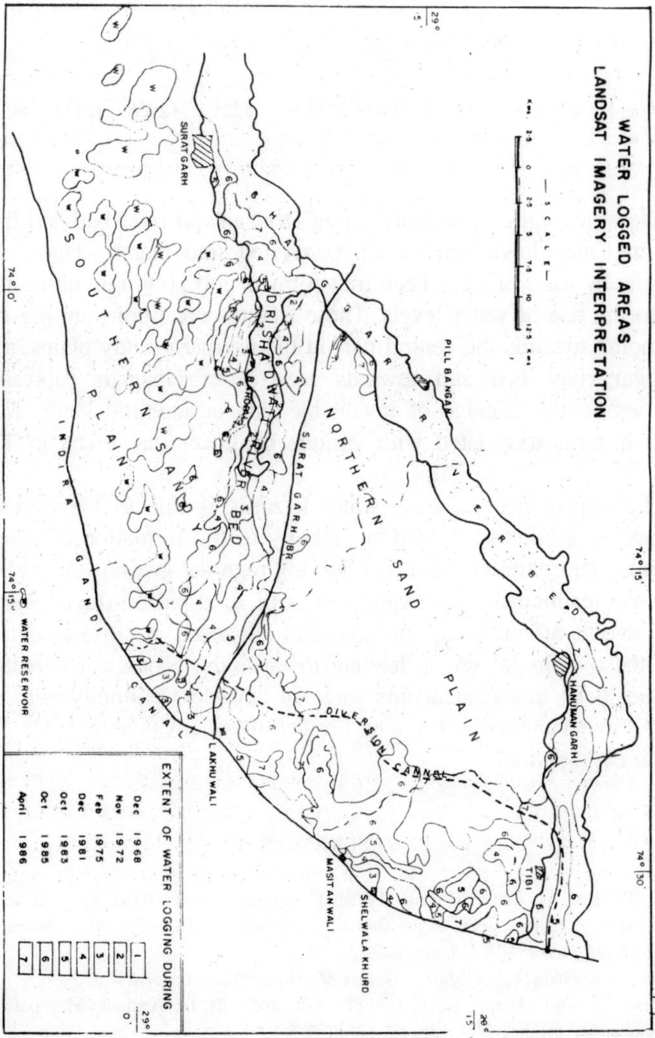

FIG. 9.1: **Water Logged Areas: Landsat Imagery Interpretation.**

TABLE 9.2: Periodical Increment of Waterlogging Conditions (Remote Sensing Data)

	Areas under waterlogging condition during year						
	1968	1972	1975	1981	1983	1985	1986
Waterlogging areas (hectares)	987	2825	5325	8225	12406	32593	44625

Once rich agriculture lands along Drishadwati river bed and Indira Gandhi Canal have either been converted into marshy lands along topography lows or have been transformed into poor agriculture lands due to the rise of water levels. These adverse conditions are presently advancing towards the central part in the northern sandy plains, along Ghaggar river bed and towards west from Ghaggar depressions resulting in the damage of invaluable rich agricultural lands. Major water hazards associated with various landforms are given in Table 9.1.

The appearance of these water hazards is due to the excessive seepage to groundwater leading initially to the formation of perched groundwater conditions and then joining regional groundwater regime. This has resulted in re-activation of once died palaeo-fluvial system. The adverse affects have appeared and advanced in close association with the old fluvial set up leading towards the ecological unbalance. This is to be managed at this stage in complete harmony with land water environmental set up to protect these invaluable lands from further degradation.

REFERENCES

Ghose, B., Kar A., Zahid, Husain (1980): Comparative role of the Arawalli and the Himalayan river systems in the fluvial sedimantation of the Rajasthan desert, *Man and Environement*, Vol. 4, pp. 8-12.

Herman, G. (1946): *The Art of Architecture of Bikaner State, Oxford*, p. 25.

Jacobson, T. and Adam, R.M. (1958): Salt and silt in Ancient Mesopotamian, *Agriculture Science*, Vol. 128, pp. 1251-58.

Man, H.S. Malhotra, S.P. and Kalla, J.C. (1974): Desert Spread — A Quantitative analysis in the arid zone of Rajasthan, *Annals of the Arid Zone*, Vol. 13, pp. 103-13.

Stein, A. (1942): A survey of ancient cities along the 'Lost' Saraswati River. *Geographical Journal*, Vol. 99, pp. 173-82.

Tessitory, L.P. (1919): *Rajasthan District Gazetteers*, Ganganager, p. 26.

10

Calcretes and Groundwater Potentiality in the Panzara Basin

S.M. BHAMARE

Introduction

DRAINAGE BASIN is a better unit for groundwater studies rather than a district or a State. The Panzara basin extends between 20° 15′ N to 21°-45′ N latitude and 73° 75′E to 76° 15′E longitude and covers 2727.82 sq km area over Deccan Plateau. Nevertheless, it bears a very insignificant part of Deccan Plateau, it may be a representative of it. The entire basin is over Deccan formation which is comprising numerous horizontal lava flows of basaltic composition. It is impervious and playing a role as an aquitard (Karlekar, 1986) for groundwater. Besides there are numerous east-west dykes along the course of river Panzara. They are acting as impartial dams for underground water; however some weaker dykes are partial to some extent. The recent alluvium is yet another important formation in relation to groundwater since it is playing both roles aquiciude as well as aqauifer. About 1/3rd part of the Panzara basin covers alluvium. Its thickness increases downstream to a maximum at mouth. Besides, there is one more formation, i.e., calcrete which is playing different role in groundwater phenomena, however it is an output factor. Calcrete is a very striking feature of the Panzara basin, which covers most part of the basin in different context.

Climate is an important input factor in a groundwater system of Panzara basin. It is very difficult to classify the Panzara basin in a specific climatic regime. The ombrothermic diagram (Fig. 10.1) for the basin is bixeric which clearly marks three seasons with prominent summer and monsoon and protracted winter. The temperature ranges between 14°C and 41°C. Under extremities it rarely exceeds above 45°C and below 5°C. The Panzara basin receives rainfall which ranges between 500 mm (at mouth) and 1100 mm (at source). The 1100 isohyet touches the source of the basin from where rainfall gradually decreases downstream (Fig. 10.1) Today there is a dearth of natural vegetation except few patches of scrubs in remotely inaccessible areas.

Efforts have been made in this paper to visualise the distributional pattern of groundwater in Panzara basin. It also aims to what extent calcrete is potential in relation to ground. The methodology included the establishment and mapping of groundwater contours on the basis of groundwater observed data with Dhule District Groundwater Survey Department. Besides, diagrammatic disposition of groundwater and calcretes have been constructed (Fig. 10.3).

Calcretes in the Region

The occurrence of calcrete is a very striking feature of geology and geomorphology of the Panzara basin. Locally, they are called as 'Bahirat' or 'Manchat'. Both the terms are used by local well diggers in the region. In connection with groundwater, the development of three basic types of calcretes viz. subareal, seepage and lacustrine calcretes could be better understood.

Subareal Calcretes

The subareal calcretes have ubiquitously occurred in the Panzara basin. Nevertheless, it is significant to note that they are more pronounced in the lower reaches of the river, since aridity increases from source to mouth. They are mostly powdery and nodular type. The $CaCO_3$ content in them ranges between 15 and 30 per cent. Their origin and development might be consequent upon groundwater fluctuations. As watertable falls during dry period, the capillary rised water evaporates out, leaving calcareous residue, i.e., subareal calcretes. The groundwater table in such type of calcrete is highly erratic, in spite of relative good aquifer in them.

Seepage Calcretes

Seepage calcretes are also ubiquitously occurred in Panzara basin. Unlike, subareal calcretes, they differs in many respects. They have

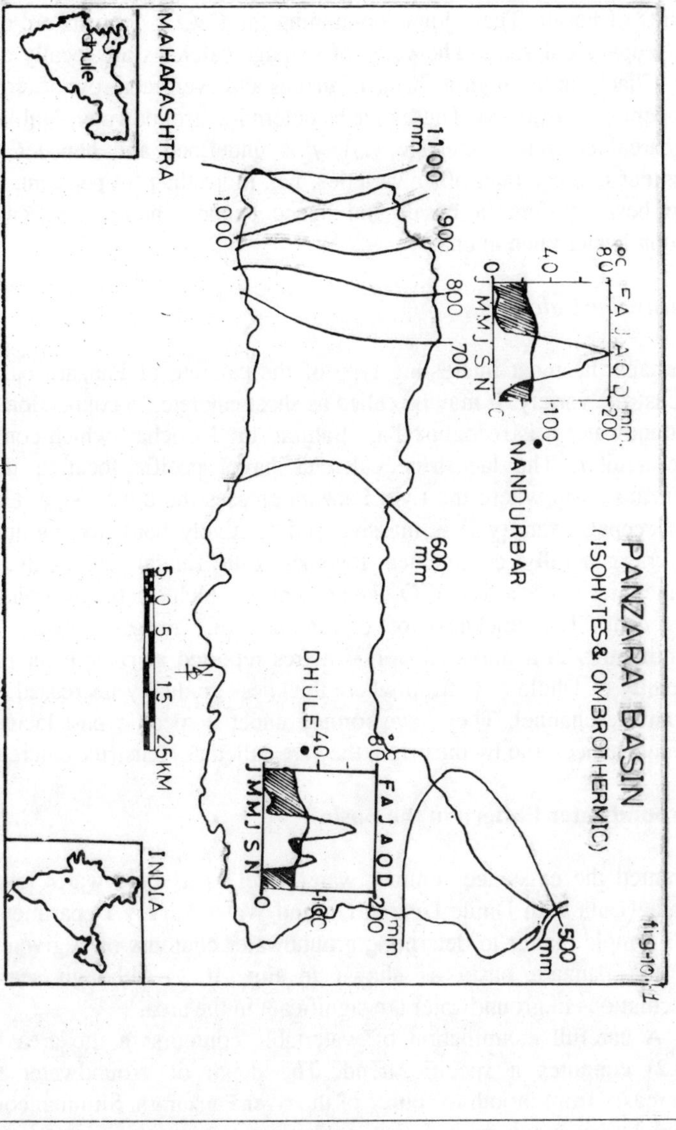

FIG. 10.1: Panzara Basin: (Isohyets & Ombrothermic.

specific location in numerous vertical as well as horizontal joints of basalt rocks. Their mode of formation is quite different than that of subareal variety. The rainwater seeps through the cracks crevices and joints. The seepage water retaining $CaCO_3$ which is precipitated in the joints of basalt. These joints containing the $CaCO_3$ deposits are called as seepage calcretes. The veins of seepage calcretes are locally called as 'Chach' in the region. "Chach" means and evergreen row of crops in rippening season too. These chachs determine groundwater. Unlike the subareal calcretes. Seepage variety is undefined and hence $CaCO_3$ content is more than other varieties, i.e., more than 80 per cent. They are best aquifers in basalt and hence become more significant in groundwater phenomena.

Lacustrine Calcretes

Perhaps the most interesting type of the calcrete of Panzara basin is lacustrine variety. It may be called as sheet calcrete. In connection with groundwater it is recognised as 'Bahirat' or 'Menchat' which confines the aquifer. The lacustrine calcretes have specific location in the Panzara basin where the river Panzara crosses the dyke range. Unlike the seepage variety it is massive and relatively hard and laminated, layers generally commented together with fluvial sediments. The cementing material is $CaCO_3$ the content of which fluctuates around 60 per cent. The thickness of calcareous crust ranges between few centimetres to a maximum of 3 metres reported at river bank in the vicinity of Dhule city, from where thickness gradually decreased away from the channel. They have formed under a specific past lacustrine circumstances and by the origin they are called as lacustrine calcrete.

Groundwater Pattern in the Basin

Granted the existence summer water level as a static water level in wells (Data with Dhule District Ground Water Survey Department), it is a simple matter to determine groundwater contours on a given map of the Panzara basin as shown in Fig. 10.2. Although seasonal fluctuations in groundwater are significant in the area.

A carefull examination of watertable contours in the area (Fig. 10.2) connotes a specific trend. The depth of groundwater table decreases from mouth to source of the river Panzarara. Simultaneously, it might be supplemented that unlike the topographic contours, the

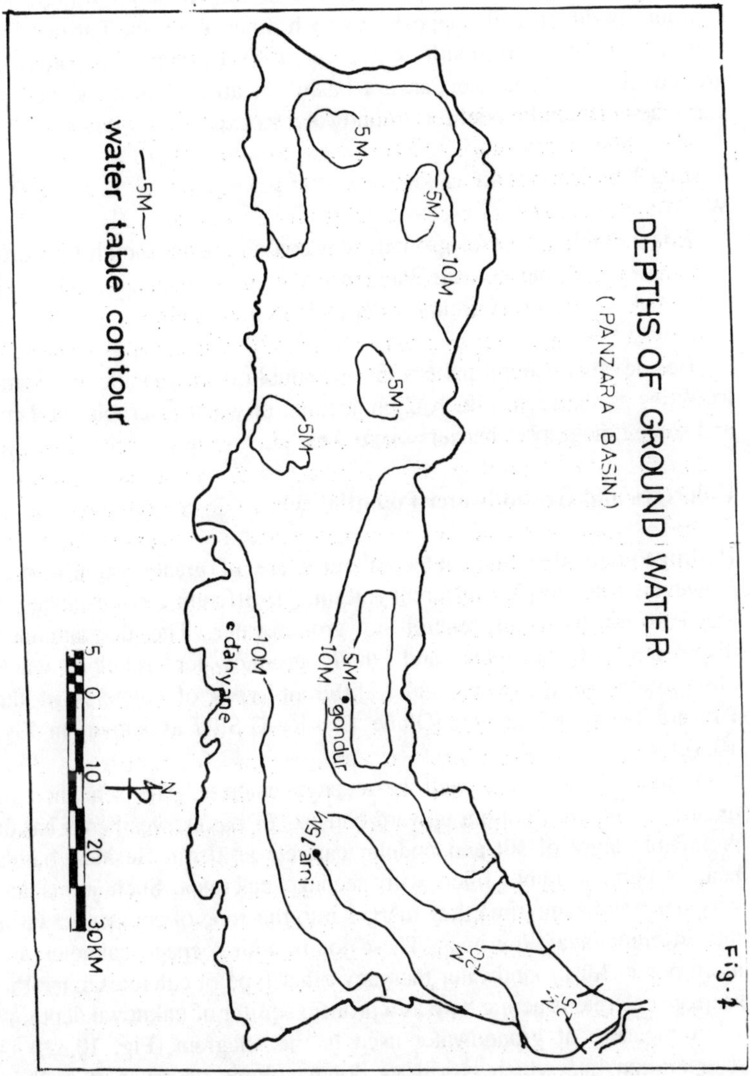

FIG. 10.2: Depths of Groundwater (Panzara Basin).

consecutive distance between the groundwater contours increases exponentially from mouth to source. It may be augmented that the spatial relation between topographic contours and groundwater contours is dimetrically opposite to each other along the Panza River from either source to mouth or *vice versa*. Considering this pattern of groundwater table in the Panzara basin, an attempt is made here to construct a general model for groundwater as expressed follows:

$$\text{Log } Y = 3.96 - 0.14X \qquad (10.1)$$

Where

Log Y = Distance of groundwater contours from mouth to source of the Panzara Basin in km.

X = Depth of groundwater table in metres.

Besides this general pattern the watertable is very shallow in some localities as shown in Fig. 10.2 on account of water reservoirs in dams and some topographic barriers like dykes.

Calcretes and Groundwater Potentiality

Hydrogeologically, area under calcretes are of greater significance. However, calcrete play different role under different circumstances in the Panzara basin in relation to groundwater. The diagrammatic disposition of calcretes and their groundwater relation water observations in the wells under different types of calcretes in the Panzara Basin. (Data with Dhule District G.S.D. as shown in Fig. 10.3).

Figure 10.3 is of a well in seepage calcrete. The sequence of material from well bottom upward consists of basalt, weathered basalt premeable layer of silt and nodular calcrete and soil. Besides, basalt bears columnar joints filled with seepage calcretes. Such joints are very deep and some time they merges into the zone of contact between the different lava flow beds. These joints with seepage calcretes are more potent for groundwater than any other type of calcretes since the seepage calcrete is acting here as a prolific aquifer of unknown depth.

A measure of groundwater used in this diagram (Fig. 10.3A) is recuperation rate which is a draw down volume of water of well in summer per unit of time. The estimated recuperation rate for the well in seepage calcrete is $9.0 m^3 h^{-1}$.

Occasionally aquifers are capped by aquiclude to prevent normal migration of groundwater and confined aquifer is formed (Easterbrook 1969). Confined aquifer are very potent in relation to groundwater and rarely occurs. However they are more common in lacustrine calcrete area of the Panzara basin which covers a considerably less area of a basin. In Fig. 10.3B the lacustrine calcrete is acting as aquiclude instead of aquifer. The weathered basalt and layer of course sand, gravel and pebble are acting as aquifer which is capped by an aquiclude, i.e., lacustrine calcrete layer. In this well the estimated recuperation rate is also 9.0m³ h⁻¹.

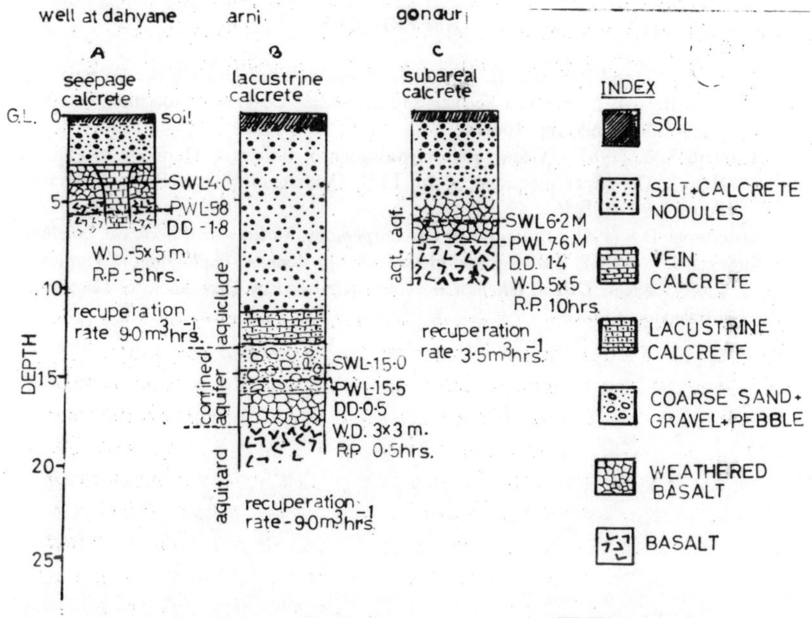

FIG. 10.3: Disposition of Calcretes & Groundwater.

Subareal calcrete covers most part of the Panzara basin and in relation to groundwater potentiality they are very poor. However water is shallower than the areas of prolific groundwater, i.e., seepage and lacustrine areas. Wells are generally dry in summer season. The recuperation rate is 3.5m³ h⁻¹ which is comparatively lower than that of wells in seepage calcretes and lacustrine calcretes (Fig. 10.3C)

Conclusion

From the foregoing discussion, it is seen that drainage basin is a better unit for groundwater studies. Considering the groundwater contours in the Panzara basin, one can conclude that the depth of groundwater decreases with exponential increase in distance from the mouth to the source of the river Panzara which can be expressed in Eq. (10.1).

Besides, calcretes have played different role in relation to groundwater in the Panzara basin. Seepage calcretes is more prolific in groundwater resources than lacustrine and subareal types; since former is an aquifer and later are aquicludes.

REFERENCES

Adyalkar P.G. and Banergee L. (1981): Hydroecology and groundwater Potentials of West Bengal" *Everyman's Science*, Indian Science Congress Association Vol. XVI (5) Oct.-Nov.-1981, pp. 140-147.

Bhamare S.M. (1986): "Geomorphic significance of Calceretic Ouricrust in Panzara Basin". The paper presented in 7th I.I.G. Conference held at Poona University Poona 7-9 Feb. 1986.

Easterbrook D.J. (1969): *Principles of Geomorphology*, McGraw-Hill Book Co. London.

Karlekar S.N. (1986): "Laterites and groundwater in Konkan" Exploration in Tropics & Professor K.R. *Dikshit felicitation volume* published by Department of Geography University of Poona.

11

Cluster Analysis in the Identification of Morpho Units: Pervikhala Drainage Basin — A Case Study

V.K. ASTHANA and KARAN VIR

A STUDY in land form geography has its own shade as compared to the traditional process-oriented study in geomorphology. The observer proceeds to analyse the visible landscape without any *priori* of denudational processes. The synthesis of observations are generally confirmed by geological observations. The aim of the study is not to confirm geological observations but to derive such geometric data, which can be readily utilised for various functions for human occupance. It is essentially descriptive and applied in approach as compared to genetic and explanatory approach of traditional geomorphological studies.

It was with the aforesaid dictum that the present authors chose to select an area of complex relief and derive some useful results. The area has already been studied by Arvind Kaur (1986) during her M.Phil. dissertation under the supervision of the senior author. She has concluded that the,

> ". . . area is under developed having faulted topography and is highly forested while the proportion of agricultural land is very small owing to hard terrain and unsuitable climatic condition."

The concluding remarks demonstrated that area has had a complex assemblage of faulted topography and has developed limitations for functional use. It, therefore, because the point of initiation which headed more detailed and careful analysis in the area and to propose certain range of landform units with rather uniform surface characteristics. Such units are referred as *morpho-units* while the attributes of each morpho units incorporate indices of various functional uses.

Traditional methods of recognising a set of morpho-units has been by superimposition of important isopleth of selected variables, and trace lines of maximum convergence to demarcate regions. Such an exercise has its own limitations specially in the selection of variables and then in important isopleths. It is only subjective appreciation of the investigator which selects important isopleths. Although, the exercise is not arbitrary, for it is guided by the density patterns of various physical attributes of the area, its contours and drainage maps, still some kind of objectivity was needed to make the exercise more precise. Another objective in mind was to assess if the recognition of morpho units drawn by traditional method stands confirmation of some kind of objective criteria. It was, therefore, decided to employ cluster analysis method of demarcating morpho-units and comparing it with the traditional method.

The Pervikhala drainage basin falls in Himachal Pradesh bounded by 30° 40′ N to 30° 55′ N and 77° 12′ E to 77° 24′ E. The Pervikhala is a tributary to the Giri river which follows a thrust plane across an area of intense folding. (Oldham, 1973) The stream runs in SE to NW direction. Its basin is formed by Salan-Rajgarh range in the south running in SE-NW direction with a maximum height of 2055 m. In the northern flank the basin is formed by Devti Dhar range running parallel to Salan-Rajgarh range. It however, has well demarcated offshoots in NE-SW directions viz. Kango range, Palu Dhar, Bathan Dhar, (Fig. 11.1). Altitude in the Devti Dhar is on an average higher with its maximum at 3000 m in the extreme NE tip. The basin has well developed ridge settlement along Salan-Rajgarh region, while the valley settlements are found along the offshoots of Devti Dhar range. The whole basin is largely forested, while small strip along valley slopes are utilised for agricultural land use. Important clusters of settlements are Rajgarh, Ratali, Bhanat, Bathan Thara. The Pervikhala stream drains due west and the Giri drains in SW directions.

Morpho Units Demarcation by Superimposition of Isopleths

First exercise in the present study has been to prepare set of distributional maps of some selected attributes. The attributes selected are Absolute Relief, Relative Relief, Slope, Dissection Index and Drainage Texture. The value of all these attributes were calculated on a one sq km grid map. To overcome the aspect of subjectivity in deciding the class intervals, z scores of each attribute was calculated ($z = x - x/\sigma$, where z is the 'z' score, score, x = individual value, x = mean of the data and σ = standard deviation). The range of 'z' score was divided into following 5 categories:

Categories	'z' Score
I	> -1.5
II	- 1.5 to - 0.5
III	+ 0.5 to - 0.5
IV	0.5 to 1.5
V	> 1.5

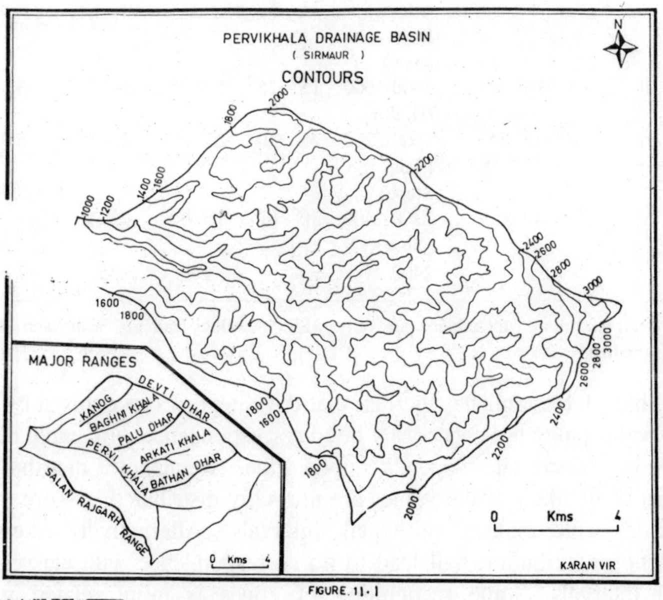

FIGURE. 11·1

FIG. 11.1: Pervikhala Drainage Basin (Sirmaur) Contours.

These categories were then smoothened in the form of isopleth maps and broad patterns noted. '*z*' score of absolute relief broadly coincided with the altitudes of 100 m, 1600 m, 200 m and 2400 m respectively. The recognition of such classes by simple altitudinal zone maps could not have been possible. The classes also broadly divided basin into five functional units of the drainage basin. Functional units have been named after Hack and Godlet (1960) with an addition of Peak, locally recognised by authors. Broad association of 'z' score and the actual values are given in Table 11.1.

TABLE 11.1: 'z' Scores and Actual values

Categories	Approximate coincidence					
	Scores	Absolute Relief (m)	Slope (degrees)	R. Relief (m)	Dissection Index	Drainage Texture (km)
I	> -1.5	1000 & below (foot slope)	<15°	< 150	< 1.1	< 1
II	-1.50 to -0.5	1000-1600 (Hollow)	15° -25°	150-250	.1-.15	1-3
III	+0.5 to -0.5	1600-2200 (Side Slope)	25° -30°	250-350	.25-.35	3-5
IV	0.5 to 1.5	2200-2400 (Nose)	30°-40°	350-500	.25-.35	5-7
V	> 1.5	2400	> 40°	> 500	>.35	> 7 km

Note: Approximate coincidence are only with 'z' values and not with each other values.

Table 11.1 establishes the fact that the categories recognised by 'z' score values may not necessarily be in correspondence with equal class intervals, since it depends upon the frequency distribution pattern of the data. If the values are normally distributed, *z* score may coincide with classes of equal intervals. Alternatively skewed frequency distribution will lead to no correspondence with classes of equal intervals. Some recognition of zones is more related with distributional patterns, it was considered that the traditional class categories be substituted with 'z' score categories. The values corresponding to similar 'z' score are, however, not inter comparable, as the distribution of each attribute differ and have different mean values. The distributional maps of 'z' score were then superimposed (Fig.11.3) and important isopleths were recognised. The basic criterion of the importance, was the association of distributional trends with the major functional units recognised on the basis of altitude. The isopleths so recognised are shown in Table 11.2.

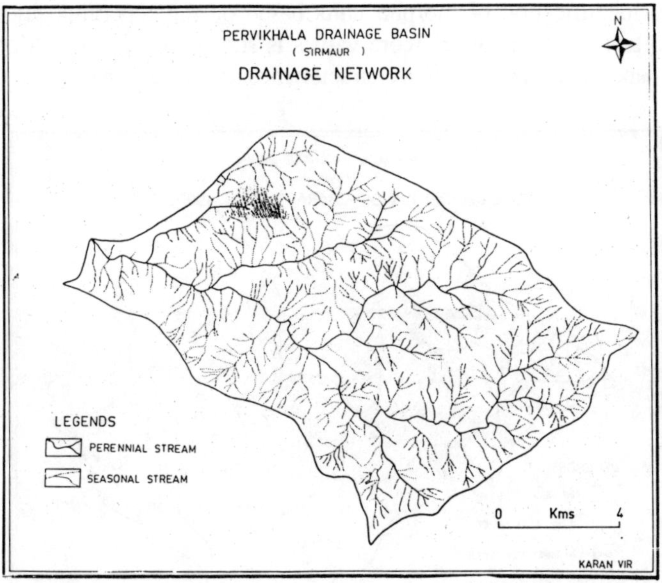

FIG. 11.2: Drainage Network.

TABLE 11.2: The Attribute and Important Isopleths

Attribute	Important Isopleths	
	High	Low
Relative Relief	350-500m	150-250m
Dissection Index	.25-.35	.1-.15
Slope	20°-40°	—
Drainage Texture	5-7km^{-1}	—

Figure 11.3 represents above listed isopleths, superimposed together. It can be observed that the *Peak* is the region of high altitude, low relative relief, low dissection index and moderately high slope. *Nose* has average characteristics. *Hollow* region has maximum convergence area of varying characteristics. This is the zone of maximum activity and associated distributional patterns. Because of this variation one

sub-region has been recognised within Hollow region in the area of maximum influence, showing high dissection index. This region of high dissection index is the region of low absolute relief and high relative relief. Such a zone is witnessed by larger stream occurrence. Final identification of morpho units based on the superimposition of important isopleths of '*z*' score values is shown in Fig. 11.4. Morpho units and their representative characteristics are listed Table 11.3.

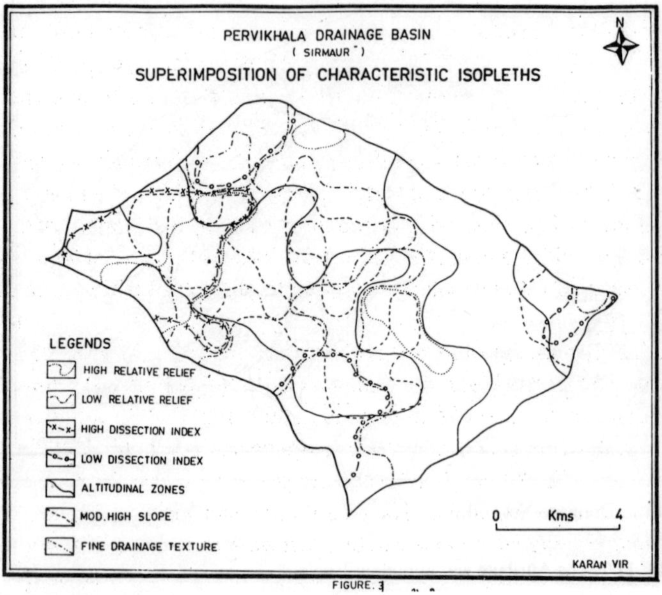

FIG. 11.3: Superimposition of Characteristic Isopleths.

TABLE 11.3: Morpho Units and Their Characteristics

Morpho units	Absolute Relief	Relative Relief	Dissection Index	Slope	Remarks
Peak	> 2400 m	> 400 m	< .15	25-30°	One subregion
Nose	22-2400 m	200-400 m	.15-.2	15-25°	
Side Slopes	16-2200 m	250-400 m	.15-.2	15-25°	Two subregions
Hollow	10-1600 m	200-250 m	.15-.2	15-25°	
Max. confluence Zone	14-1600 m	400-600 m	.25-.4	25-30°	
Foot slope	< 1000 m	300 m & below	< .25	15-25°	

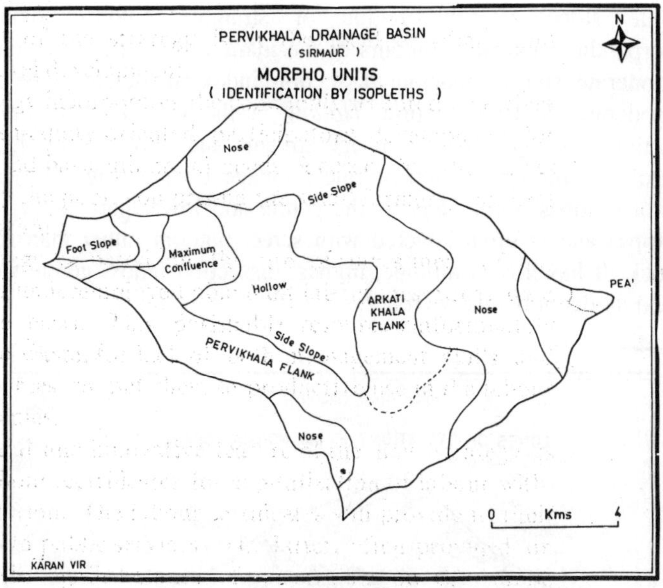

FIG. 11.4: Morpho Units (Identification by Isopleths).

Besides, the listed morpho units in Table 11.3, two sub-regions are identified in the side slopes and peak. The identification of Pervikhala and Arkatikhala flanks is based on relatively moderate slope in the Pervikhala flank as compared to Arkatikhala flank, i.e., 15-25° v/s 25-30°. Similarly, the area bounded by dotted line in the peak region is the zone of low relative relief, i.e., 150-250 m as compared to 350-500 m.

Qualitative description of various attributes in different morpho units can be recognised as follows:

Peak is the region of very high altitude (> 2500m), moderate to high slope (25-30°) and low dissection index (<.15). *Nose* is more or less similar to peak region, but has rather low altitude and gentler slope with a higher dissection index. The region, being traversed by higher order stream, has developed higher dissection. In a normally valley, occurrence of nose is around the periphery of the basin having convex slope, and coarse drainage texture. It is broken at two places which could be scars of drainage reversals. Side slope are the region of straight contours, running perpendicular to the contours of the main valley. It is a region of moderate slope, moderate dissection index, moderate slope and moderate drainage texture. *Hollow* is the region of maximum activity. It receives all drainage streams and develops a concave profile. This is a region of fine drainage texture (Fig. 11.5) and moist slopes. *Foot Slope* is the gentle and lower part of the side slopes and is often covered with scree material rather than solid rock. It has lowest altitude, highest dissection index, gentle slope and moderately relative relief.

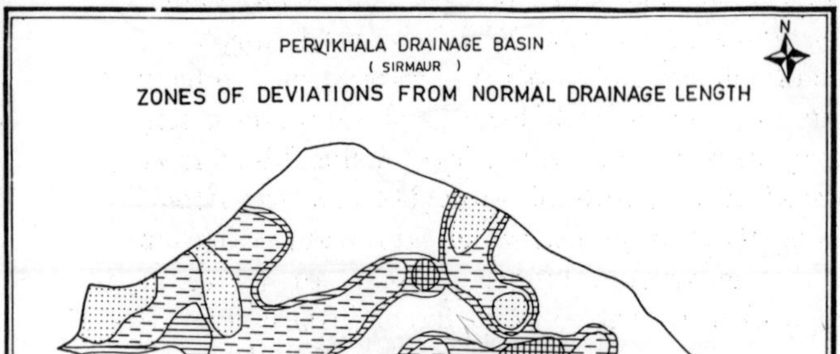

FIG. 11.5: Zones of Deviations from Normal Drainage Length.

Figure 11.5 exhibits a pattern of drainage texture, which has been generalised by following slopes:

(i) Only perennial stream length is counted rather than total length, with an assumption that the streams flowing all the year round are most potent agent of denudation as compared to seasonal stream.

(ii) Perennial length of streams per unit area is substracted by a constant of 1.4 km. This constant is the threshold value of water erosion according to Leopold *et al.* (1969). On an average one sq km area is sufficient to produce a drainage length of 1.4 km. Any drainage length higher than the value will be moist slope and alternatively the regions having smaller drainage length be a region of dry slopes.

One of the two steps of generalisation, the pattern of drainage length shows interesting pattern in Fig. 11.5. The flanks and peaks have average required drainage length of 1.4 km but the hollow region has both moist slopes and dry slopes. The occurrence of dry slopes is intriguing. Cooke *et al.* (1924) have described that the Nose area is generally the domain of dry slopes, since it receives least moisture from the upland area, while it provides shedding distance to the formation of drainage streams. An examination of contour map (Fig. 11.1) and the drainage map (Fig. 11.2) reveals that the patches of dry slopes lie along water divide between Pervikhala and Arkatikhala streams, mid-channel way of Pervikhala and the foot slopes of the Pervikhala. The pattern of drainage length in the eastern half seems natural, if we assume that the original drainage was due SW draining into Giri river flowing in NW to SE direction. The region seems to be the area of truncated stream, which has not readjusted its stream length because of time lag. The plausible theory to account for this anomalous occurrence is that original drainage alignment was due NE to SW, formed by upper tracks of the Pervikhala and Arkatikhála. These tributaries were drainage into the Giri river flowing further southwards. Tectonic disturbances resulted faults along Giri and Pervikhala drainage line in NW-SE direction. There was a general upthrust in the NE part of the study area and truncation of streams which now started flowing in SE to NW direction. The catchment of Arkatikhala tributary, therefore, appears more normally developed as compared to the lower section of Pervikhala stream. This assumption is further supported by the occurrence of highest altitude in the NE tip of the basin, and the gaps in the nose unit of Pervikhala drainage basin. However, it needs further detailed observations to finally check the validity of the assumption.

The identification of morpho units not only helps to divide regions into sub-units of broadly uniform character, but also in the understanding of geomorphic history, as revealed in the above mentioned text.

Next part of the exercise was to recognise morpho units by cluster analysis which is primarily a computer-oriented exercise. Use of computer enhances the objectivity and the personal prejudices are reduced.

Morpho Units Demarcated by Cluster Analysis

Clustering according to Mather (1976) "has the aim of ordering objects into descrete sets". The need of cluster analysis is aptly described by him, while starting that, ". . . if one's aim is to consider simultaneously many characteristics of a large number of objects then some form of automatic selection procedure becomes essential". In fact, many methods of classification have demonstrated that they have greater failings of discrimination and subjectivity and hence a technique of greater objective procedure is searched in the cluster analysis.

Cluster analysis is generally of two kinds, i.e. (i) divisive or agglomerative and (ii) hierarchical or nucleated. In agglomerative procedure one beings with a set of individuals and then builds up groups by process of accumulation, whereas the reverse procedure is followed in divisive methods. Agglomerative methods allow the simultaneous consideration of all characteristics of the objects. It is due to these qualities that the present authors chose to select agglomerative method of cluster analysis. The data of all five attributes, i.e., absolute relief, relative relief, dissection index, slope and drainage texture was arranged with respect to each grid in the study area. The grids were then serially arranged from 1-113 with its key of location. The data was then fed to computer with an instruction to search unlimited centroids. It finally searched 3 centroids, as shown in Table 11.4.

It apparently exhibit that the degree of generalisation has increased in this method and only 3 clusters, instead of 6 functional units are searched. Broad equivalent of these clusters are mentioned in the column of remarks. These clusters have 35, 40 and 38 members respectively. Figure 11.6 shows an idealised demarcation of morpho units based on cluster analysis. Cluster 1 broadly confirms with hollow and foot slope, 2 with side slopes and 3 with peak and nose. The pattern of second cluster is rather peculiar. It has the coarser drainage texture, moderate slope, high dissection index and high relative relief. The zone, although, demarcates the area of maximum activity, but coarse drainage texture suggests that the drainage length has still not adjusted with the topography. This imbalance points out the fact that the basin is truncated and has witnessed changes in the base level. Earlier assumption of NE-SW flowing streams parallel to Akratikhala is also supported by this pattern. Another important feature is the distribution of side slopes. It demarcates areas of high altitude. Low relative relief, moderate dissection index and fine drainage texture. The alignment of this cluster is along the flanks of Baghni Ka Khala and the upper catchment of Pervikhala. Baghni Ka Khala appears to be the most recent stream formation initiated by the change of base level near foot slope, while the upper catchment of Pervikhala has experienced drainage diversion due to NW-SE trending fault.

TABLE 11.4: The Data of all Five Attributes

Category	Absolute Relief (m)	Relative Relief (m)	Dissection Index	Slope (Degrees)	Drainage Texture km^{-1}	Remarks
1.	1870.3	230.3	.1	23.6	4.3	Hollow & Foot slope
2.	1804.7	436.5	.2	23.9	3.7	Side slopes
3.	1995.2	251.4	.1	15.8	6.6	Peak & Nose
x (Mean)	1896.9	323.5	.17	24.5	5.6	—
σ (Standard deviation)	346.5	120.6	.06	7.5	1.9	—

Broadly the demarcations of morpho units by cluster analysis and by superimposition method correspond within permissible range of variation. However, the greater degree of generalisation in cluster analysis has led to rather large simplication of information, and overcasts the details of topographic fluctuations. It has, however, helped in confirming the general alignment of morpho units recognised by superimposition of important isopleths.

Authors, opine that the method of demarcating morpho units of superimposition has its own viability in the analysis of an individual drainage basin, while the cluster analysis has its advantage in broad regionalisation of a large tract including many basins and related informations. For small basin, cluster analysis has some limitations, as it overcasts many fluctuations of topographic data.

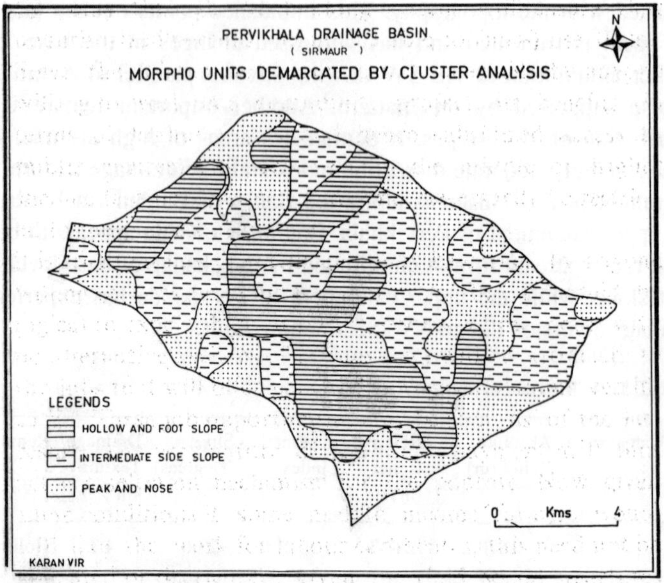

FIG. 11.6: Morpho Units demarcated by Cluster Analysis.

Table 11.5, gives a comparative picture of functional units and cluster units. Peak and zone of maximum confluence more or less completely coincides with second unit of cluster. Similarly, hollow and foot slopes coincides with first unit and Nose with third unit. There is more or less normal distribution of cluster units in side slopes with its mean along second unit of clusters.

Table 11.5 has slight variations with the qualification of cluster units in Fig. 11.6. The terms mentioned in Fig. 11.5 with each cluster unit are as under:

Cluster 1 : Hollow & Foot slope
Cluster 2 : Intermediate Side slope
Cluster 3 : Peak & Nose

The variation is with regard to occurrence of 66 per cent 2nd cluster in peak region, which is termed qualitatively with side slopes. In fact, the region of peak and Nose if taken together, highlights the importance of

3rd cluster, as against 2nd cluster. Occurrence of 2nd cluster is significant in the peak region only. Hollow and Foot slopes more or less show large correspondence with cluster unit No. 1. side slope unit also show larger correspondence with 2nd cluster unit. The characters of these clusters are already mentioned in Table 11.4.

TABLE 11.5: Comparative Picture of Functional units and Cluster units

Functional units	Occurrence of Clusters 1 no	1 %	2 no	2 %	3 no	3 %	Total Occurrence no	Total Occurrence %	Correspondence with cluster
1. Peak	-	-	4	*66*	2	34	6	100	2
2. Nose	4	12	11	35	16	*53*	31	100	3
3. Side slopes	15	33	19	*43*	11	24	45	100	2
4. Hollow	14	*70*	3	15	3	15	20	100	1
5. Zone of Maximum Confluence	1	16	5	*84*	-	-	6	100	2
6. Foot Slopes	3	*60*	1	20	1	20	5	100	1
Total	37	32	43	39	33	29	113	100	-

Correspondence of cluster units in percentage is underlined.

To summarise the findings of the present investigation, author: hold that the method of cluster analysis can only be an aid tc interpretation and not an ultimate. The importance of superimpositior of '*z*' score values of important isopleths has its distinctive merit which not only brings out greater variety of morphological units bu also unravels geomorphic history of the area. The inferences drawn however, are further subject to detailed field observations, an geological findings. Detailed remote sensing data could als supplement the findings.

REFERENCES

Cooke, R.V. and Doornkamp, J.C. (1974): *Geomorphology in Environmetal management*, Clarendon Press p.13

Hack, J.T. and Goodlet J.C. (1960): Geomorphology and Forest Ecology a mountain Region in the Central Applachians, US. Geolburn Prof. Paper 347, pp. 66.

Kaur, Arvind (1986): Analysis of Physical Environment of Pervikhala Drainage basin: M.Phil. dissertation August, Deptt. of Geography, Kurukshetra University,

Kurukshetra, (unpublished).

Leopold, L.B. *et al.* (1980): *Flurial Processes in Geomorphology*, Freeman & Co. 1964, reprinted S. Chand & Co. New Delhi, p. 146.

athur, P. (1976) : *Computational Methods of Multivariable Analysis in Physical Geography*, John Willy & Sons, p. 309.

Oldham, R.D. (ed. & rev.) (1973): A *manual of the Geology of India and Burma*; ed. & revised by R.D. Oldham, GSI publication 1973, p. 452, Vol. 1.

12

Occurrence of Sandstone in the Tapi Floodplain

Y.V. PATIL

Introduction

THE TAPI river extends between 74°E to 78°E longitude and 20°N to 21°N latitude. The river Tapi itself is a rift valley and it's trough is filled with consideration deposits of alluvium. The alluvium layers are much thicker and wider. These alluvium layers are composed of yellow and brownish clays with gravels and *kankars*. A compact hard layer of sandstone is observed at number of places in the Tapi, and it's tributaries, like Panzara and Girana, i.e., in Tapi near Bhusawal, in Panzara near Dhulia, Betawad and in Girana near Pachora and Jalgaon. The thickness and other characteristics of sandstone varies from one location to another. The sandstones of the maximum thickness are observed near Bhalsiv a place 20 km away from the Bhusawal. The region is intensively eroded by gullies and ravines, and is known as "Yaval ravines".

A sandstone is a sedimentary rock made up of grains of sand kept together by silt, clay and precipitates of mineral cement. The present study of sandstone centres on the framework character and voids and void filler. The sandstone is observed at different location but they are principally located on the northern bank of the river. The observations of the well sections located on the alluvial deposit gives clear idea about their thickness and occurrence. Their thickness varies considerably from few inches to few tens of feet.

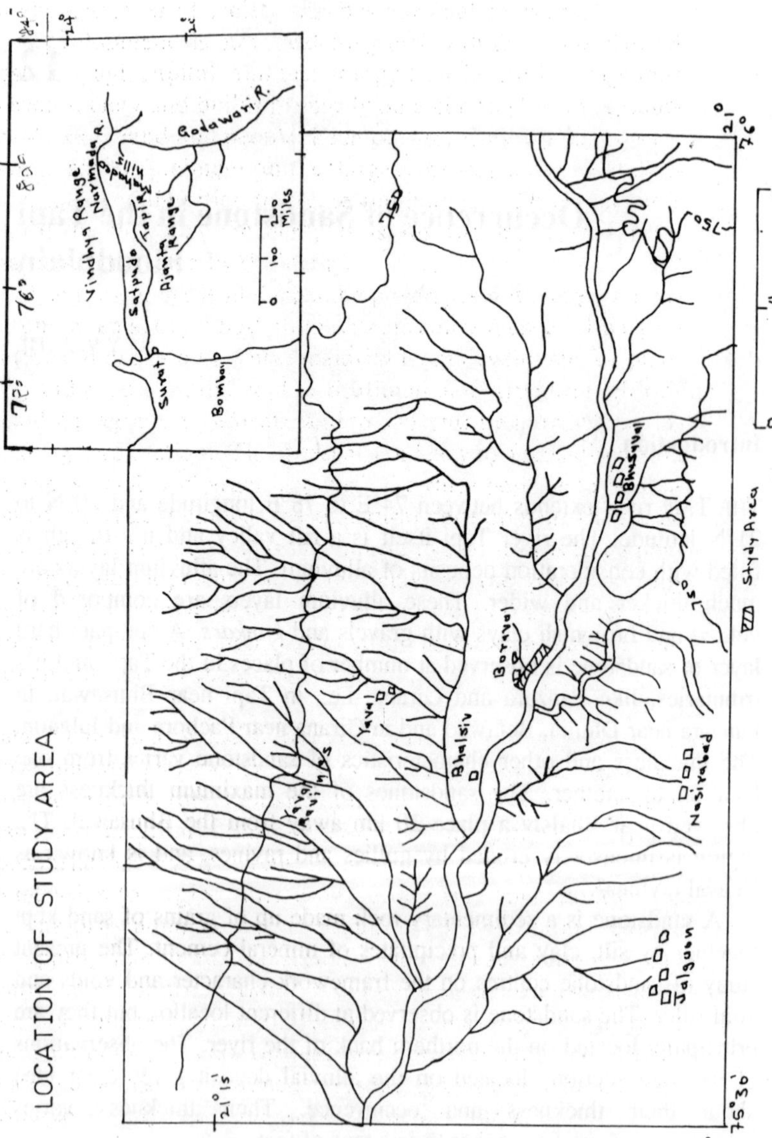

FIG. 12.1: Location of Study Area.

Methodology

The present study is based on the detailed fieldwork carried out in number of visits, data obtained from the topographical maps and the analysis of the material. The work has been divided into two major parts.

1. Fieldwork
2. Analytical work.

The analytical works includes chemical analysis and physical analysis of the sandstones. In profile analysis of the sandstones, generalised sandstone profile has been drawn to account for the general idea of the material and depositional processes. During the field visits photographs and cross-sections has been drawn to get the clear idea about the sandstones.

Generalised Sandstone Profile

For the detailed study of sandstone a cross-section is studied which is located near a village Bhalsiv, a place on the bank of the Tapti river. At this location there are number of sandstones boulders in the river channel. They are hard compact and looks like a concrete blocks, and cannot be broken by the hammer.

These sandstones show some remarkable layering and there is usually a clear difference between that part of the sandstone near the surface and that laying several feet below.

In a sandstone profile seven main horizons are to be seen. Their description and composition (Table 12.1) is discussed as below:

TABLE 12.1: Layers of Seven Main Horizons

Layer No.	% of calcium carbonate	% of sand	% of silt clay & cementing material	Characteristics
1.	10	5	85	Yellow alluvium with *kankars* & sand.
2.	10	50	40	Loose friable sandstone.
3.	25	65	10	Compact massive sandstone, difficult to break by hammer.
4 & 7.	10	5	85	Loose clay fine texture.
5.	20	60	20	Sandstone of very fine sand, hard & massive.
6.	–	–	–	Pebbles of different size only.

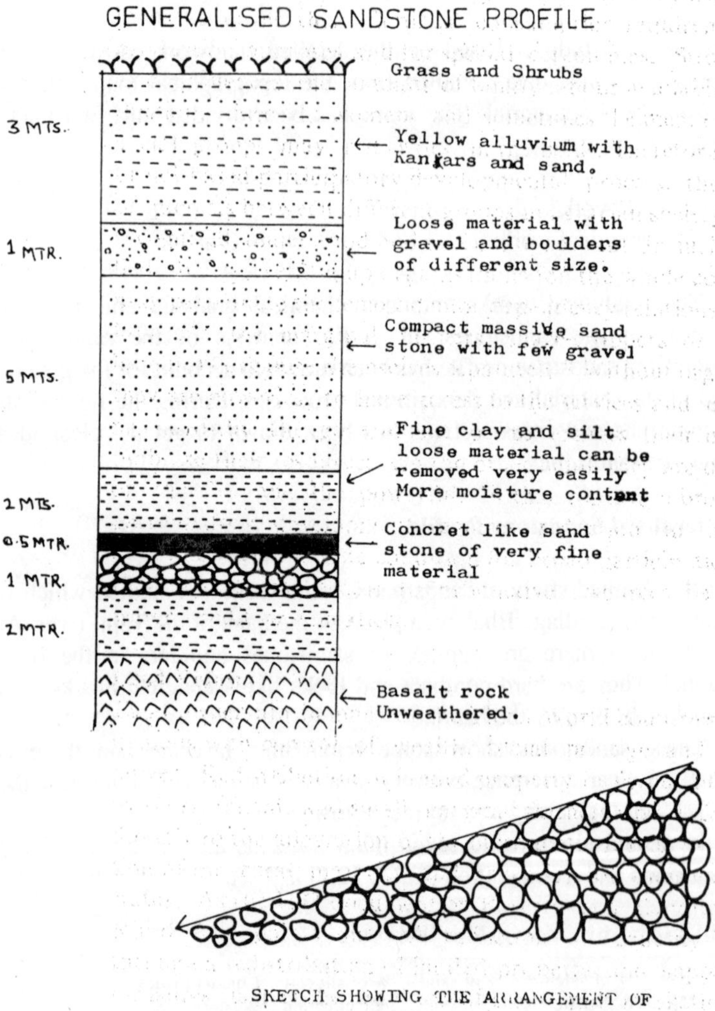

SKETCH SHOWING THE ARRANGEMENT OF PEBBLES AND THICKNESS OF PEBBLES

FIG. 12.2: Generalised Sandstone Profile.

The upper most layer consisting of the yellow alluvium with *kankars* and sands. The thickness of this zone is 10 feet thick. It is covered by the small grasses and shrubs. This layer consist of 10 per cent of the calcium carbonate.

The second layer consists of loose material, with gravels and boulders of 3 feet thickness. This zone has friable and loose material which can be easily removed by hammer. This layer consists of 60 per cent of the sand, 10 per cent of the calcium carbonate. These first two layers can be called as the *zone of alluviation or zone of leaching*, from which the cementing material and calcium carbonates are rapidly moved downwards and loss of cementing material and mineral salts takes place. This leaching of the precipitates of calcium carbonate and mineral cements makes the layer friable, loose and porous.

The third layer is a layer of hard compact massive sandstone of 15 feet thickness, the proportion of calcium carbonate suddenly increases up to 25 per cent. The sand particles are highly cemented by precipitates of calcium carbonate and other cementing materials. The sandstone of this layer cannot be broken by the hammer, and the hammer bounces from the sandstone. The fourth layer consist of fine silt and clay, because of its fine texture the clay does not permits the downward movement of precipitates of cementing material, the downward movement of precipitates is nearly absent so that the material gets accumulated in the third layer, which can be known as a zone of illuviation or zone of accumulation. The voids are completely filled by the precipitates of the cementing material.

The fourth layer consist of fine silt and clay of 5 feet thickness, because of the fine texture it has more moisture content than third layer, at some places, the removal of clay from this zone gives appearance like a caves in the study area.

The fifth layer consists of a cementing of the very fine sand of one feet thickness. It is a concrete-like hard structure of very fine sand consisting 20 per cent of the calcium carbonate. Above this layer there is a layer of silt and clay which permits downward movement of very fine material only giving rise to the formation of this layer.

The sixth layer is a layer of gravels and pebbles, without any cementing material. The average thickness of this layer is 5 feet, but it decreases in the downward direction. Below this layer i.e., seventh layer there is a layer of fine silt and clay similar to that of fourth layer. Below this seventh layer there is a hard basalt rock located at the level of the river bed.

The deposition of the alluvium and formation of the sandstone on the northern banks, supports to that the present entrance of the Tapi in to Jalgaon is said to be a new channel, it is possible that the river once flowed for the north along deep gully which may still be seen and joined the present channel near the village Ainpur about 20 miles east of Bhusawal.*

Conclusion

1. The formation of sandstone indicates that the source area and site of deposition be tectonically very stable or that the sand has gone through the several cycles of sedimentation.
2. The depositional processes are physical but the formation of sandstone by the processes of leaching is a chemical process through which the precipitates of the cementing material moves downward direction and their accumulation gives hardness to the zone of accumulation.
3. The period of formation of loose friable material, and it's deposition represents pluvial climate whereas the sandstone formation, i.e., lithification represents the semi-arid, arid climate.

REFERENCES

1. *Gazetteer of Jalgaon District*, 1880.
2. F.J. Pettijohn (1969) *Selimentary Rocks.*

---

* *Gazetteer of Jalgaon District*, 1880.

A Geomorphological Investigation of Rock Weathering and Soil in the Mawsynram Area of East Khasi Hills District, Meghalaya

GIRISH CHANDRA PANDA and R.K. RAI

GEOMORPHOLOGICAL INVESTIGATION of weathering of rocks and soil formation is an important developing field of research. The distribution and thickness of the weathering products, or more properly the absence of soils are the prime importance in determining the suitability of any given area for various landuses. The absence of soil places constraints on agriculture, it increases the costs and engineering problems encountered in the preparation of agricultural field and also in construction of irrigation canals, tanks and wells. These geomorphic factors become more crucial as the population increases and forestry become more highly organised and commercialised.

Therefore, the study of weathering of rocks along with acquaintance of certain soil characteristics are considered to be fundamental to any enquiry into contemporary geomorphological research. It is in this perspective the present study has been undertaken.

Study Area

The area under study lies between 24°57′ to 25°27′ N latitudes and 91°26′ to 91°43′ E longitudes, which covers a territorial extension of 454.4 km². The diverse geology and geomorphic characteristics, i.e., the occurrence of the undulating crystline high plateau with numerous residual hills to the north, gently dipping sandstone table land in the centre and low-lying dissected terrain in the south present a variety of landform features (Panda, 1987). One phase of evolutionary history of its geological landscape is one of the recurring uplift and down sinking during the past more than 100 million years, which led to the formation of the present spectacular landscape geometry (Bandopadhyaya, 1972). The variety of sediments from Archaean to the recent periods forming the region is discernable from its protruding surfaces. Its geological chronology seems to have similar sequence records as that of the Chota Nagpur Plateau and the Rajmahal Highlands (Rai, 1986). In terms of climate the study area can be put under tropical monsoon climate, with an average rainfall about 12,000 mm annually. Thus, the study area is providing an ideal geological and environmental base, which favours weathering and decomposition of bed rocks of different ages.

Methodology

In this paper an attempt is being made to examine the spatial nature of weathering variations on different bed rocks with an inventory of soil characteristics and its relationships with toposequence and vegetation.

The material presented here is based primarily on the field observations and bed rock weathering, vegetation, and measurement of soil depth, slope and altitude in various parts of the Mawsynram area, wherein road cuts and other excavation have been considered in detail. The Rougerie guide (Thomas, 1974) to the study of rock weathering for field observation have been considered to examine the general nature of weathering on different bed rocks. Data concerning soil characteristics, slope, altitude and vegetation were obtained for each selected sites (Fig. 13.1) at fixed intervals by employing standard methods (Smith and Atkinson, 1975).

Results and Discussions

Weathering on the Archaean and Pre-Cambrian Rocks

These bed rocks are comprised of granitic and metamorphic rocks. On relatively steep slopes around the village Nongshymit most granitic rocks tend to weather uniformly to gruss, separated by scattered outcrops and tors. Considering the Rougerie guide of rock weathering it is found that the pink granites of the study area most susceptible to gruss alternation. The granite gneisses containing quartz is intermediate in susceptibility; and the porphyritic granite is seen around villages Nonpung and Sinei along with tourmaline granite near village Umblei are least susceptible. The metamorphic rocks of mica-schists are quite resistant to weathering. It has been observed that 'slacking' is the dominating weathering process of this area. This process has given rise to accumulation of granite blocks (Tors) in many parts of the study area. Table 13.1 shows that the soil depth and soil characteristics vary greatly according to topographic position on granite bed rock surfaces near the Tyrsad village sites.

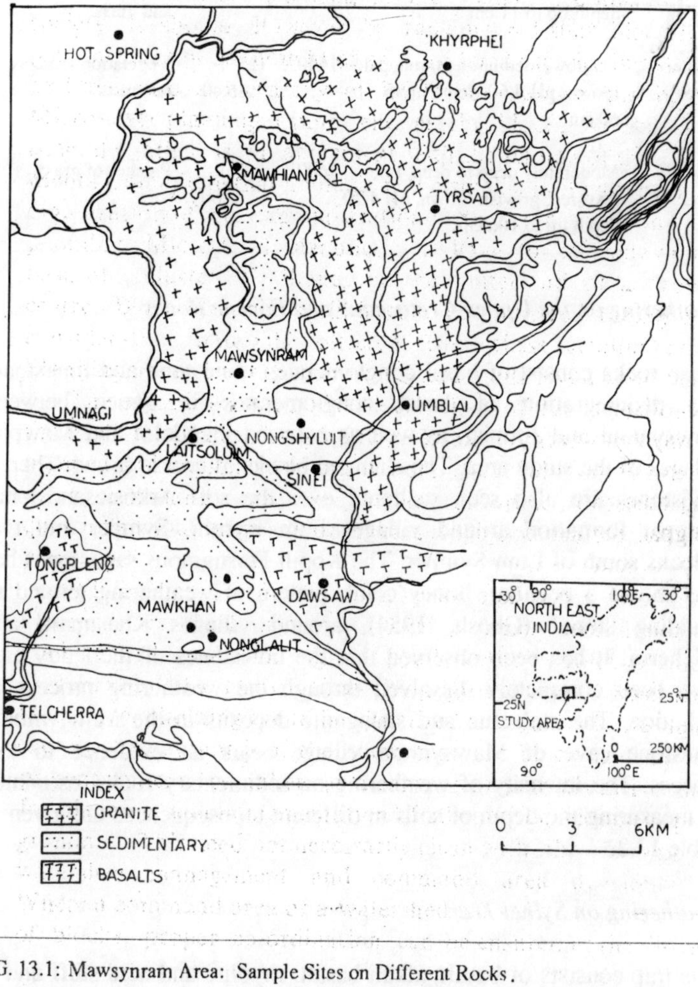

FIG. 13.1: Mawsynram Area: Sample Sites on Different Rocks.

TABLE 13.1: Sample Area Surveyed at the Tyrsad Village Site of the Mawsynram Region–Meghalaya

Topographic Location	Soils	Mean elevation in m.	Mean slope in %	Dominant Vegetation
Crests	Very shallow, coarse textured, dark greyish brown, pH 4.50, depth up to 11 cm	1700	(5°) 8.75	Relect vegetation stunted pine, shrubs and grass.
Middle Slopes	..allow, medium textured, pale brown, pH 4.60, depth up to 60 cm.	1640	(25°) 36.40	Vegetation same as on crests
Valley floors	Moderately deep to deep, fine textured, greyish brown, pH 4.70, depth up to 106 cm.	1600	(7°) 17.60	Patches of grass

Weathering on the Upper-Cretaceous and Eocene Rocks

These rocks consist of basal conglomerates, sandstone and limestone. The disintegration of basal conglomerates is notice between Mawsynram and Jimpiat, as well as between Pamdoloi and Mawphu villages of the study area. The slumped blocks of coarse brown Therria sandstones are also seen scattered over the soft Arkosicsandstones Langpar formation around villages Lum Kyrphi, Symper and two hillocks south of Lum Symper. The Kopili Ferruginous sandstones has also shown a peculiar, honey comb pattern of weathering termed as 'pudding stone' (Ghosh, 1954), around villages Khasimara and Telcherra. It has been observed that the limestones of the study area have been completely dissolved through the weathering process of hydration. The stalactite and stalagmite deposits in the Mawjymbuin limestone cave of Mawsynram village bears an evidence to this process. The intensity of weathering on sedimentary rocks, estimated by measuring the depth of soils in different toposequences are given in Table 13.2.

Weathering on Sylhet Trap

This trap consists of basalt, alkali basalt, rhyolite and acid stuff and is seen along the southern part of the study area in a narrow east-west patch. The crest of this rock has been weathered in some places exposing the trap rocks as inliers. The inliers are seen at south of the village Sinei. This dolerite is greenish-black, hard and is weathered into spheroidal boulders. The final product of basalt and dolerites is a rich dark colour soil layer. The measurement of soil depth at different toposequences near Trongpliang village sites (Table 13.3) supports the above statement and confirms the high rate of weathering on the basalt bed rock surfaces.

TABLE 13.2: Sample Area Surveyed at the Mawsynram Village Sites of the Mawsynram Region, Meghalaya

Topographic Location	Soils	Mean elevation in m	Mean slope in %	Dominant Vegetation
Crests	Very shallow, coarse textured, grey soil, pH 4.50, depth up to 8 cm.	1383	(8°) 14.00	Shrubs dominated with grass.
Middle Slopes	Shallow, medium textured, light brownish-grey pH-4.60, depth up to 35 cm.	1363	(20°) 36.40	Semi-natural vegetation and other under-storey trees with shrubs.
Valley floors	Shallow to moderately deep, fine texture, very pale brown, pH 5.5, depth up to 165 cm.	1300	(10°) 17.60	Grass up to 60 cm high.

Conclusion

The study of weathering of rocks and soil characteristics at a number of sites in Mawsynram area permits a few general observations. In the first place, there is a marked variation in the nature of weathering processes on the granitic rocks, sedimentary rocks and basalts of the study area. The structure and minerological characteristics of these rocks have contributed to the formation of various landform features.

Further, great irregularity of the soil depth and soil characteristics on various rock surfaces at different toposequences are characteristic features of most of the areas. This irregularity can also be seen to consists of different vegetation covers. In all cases a correspondence exists between the susceptibility of rocks to weathering processes and depth of the soil. On the other hand, each individual site with their uniform rock formations also show a progressive thickening of soil depth from hill top locations towards valley bottoms. A closer examination of the distribution of soil colour, soil reaction (pH) and soil texture from crests to valley floors in the selected sample village sites show that on the rock basements, pedogenic evolution directly depends upon the topographical position of parent materials. This may account for the presence of red loam soils on the crest slopes as well as comparatively youthful yellow and red soils on middle slopes and valley floors.

TABLE 13.3: Sample Area Surveyed at the Trongpliang Vilalge Sites of the Mawsynram Region, Meghalaya

Topographic Location	Soils	Mean elevation in m	Mean slope in %	Dominant Vegetation
Crests	Shallow moderately deep, medium to fine textured, yellow to red dark greyish brown pH 4.85, depth up to 20 cm	950	(10°) 17.60	Natural regrowth of evergreen rainforest patches after slash and burn agriculture (*jhum*)
Middle slopes	Moderately deep, fine textured, yellow to red greyish-brown, pH 5.80 depth up to 48 cm	890	(60°) 173.20	Vegetation same as crests but sparse
Valley floors	Moderately deep to fine textured, deep, yellow to red greyish-brown, pH 5.25, depth up to 156 cm	850	(20°) 30.40	Vegetation same as the crests and middle slopes but more densely covered.

Thus, the bio-physical information obtained from each sample site do establish a close relationship between the parent materials, rock weathering, soils, toposequences and vegetation. These primary environmental factors are essential for the consideration of developmental projects in any area. Moreover, an integrated picture provides the appropriate framework for land use planning in Mawsynram region one of the least agriculturally developed and most ecologically vulnerable areas of the East Khasi Hills district of Meghalaya.

REFERENCES

Bandopadhyaya, M.K. (1972): Geomorphological characteristics of southern part of the Khasi Hills, *Geographical Review of India*, Vol. 34.

Ghosh, A.M.N. (1954): *General Report, Geological Survey of India, Assam Circle*, Shillong, India.

Munseel Soil Colour Charts (1975): *Munseel color Macbeth Division of Kollmorgess* Corporation, Baltimore, Maryland, 21218, U.S.A.

Panda, G.C. (1987): Geomorphology and Agricultural Landuse Capability Classification in Mawsynram Region, Meghalaya (unpublished Ph.D. thesis) Department of Geography, N.E.H.U., Shillong, India.

Rai, R.K. (1986): Evidences of Rejuvenation of the Deccan Foreland, India, with particular reference to the Meghalaya Plateau; *International Geomorphology*, Part II, ed. by V. Gardiner, John Wiley and Sons Ltd., London.

Smith, R.T. and Atkinson, K. (1975): *Techniques in Pedology: A handbook for Environmental and Resource Studies*, Paul Eleck Ltd., London.

Thomas, M.F. (1974): *Tropical Geomorphology : A Study of Weathering and Landform Development in Warm Climate*, Pub. McMillan Press Ltd., London.

14

Drainage, Lithology and Structure in the North-western Part of Rajmahal Hills, Bihar, India

SRINIVAS MADABHUSHI

Introduction

THE STUDY area comprises 1,300 sq km on the north-western part of Rajmahal hills between 25° 0′ N and 25° 15′ N latitudes 87° 15′ E and 87° 42′ 30″E longitudes (Fig. 14.1).

Previous Work

The geomorphological studies in the area can be traced back to James Rennell (1781) whose maps are used for a study of change in the river courses. Major Colebrooke (1803) studied the course of the Ganges through Bengal. Morgan (1959) analysed the neo-tectonic activity in Bengal basin. The geomorphological evolution of Rajmahal hills was discussed by Singh and Singh (1960). Madabhushi (1985) summarised the previous studies.

Madabhushi (1980) mapped the alluvial covered North Rajmahal coalfield with the help of 1″=1 mile aerial photographs, demarcated the possible Barakar-Pre-Cambrian boundary below the alluvial cover and prepared the lineament map and drainage map on the base 1″ = 1mile scale. The morphometric analysis was done on the 1″ = 1 mile scale maps of which the Fig. 14.2 is a reduced version.

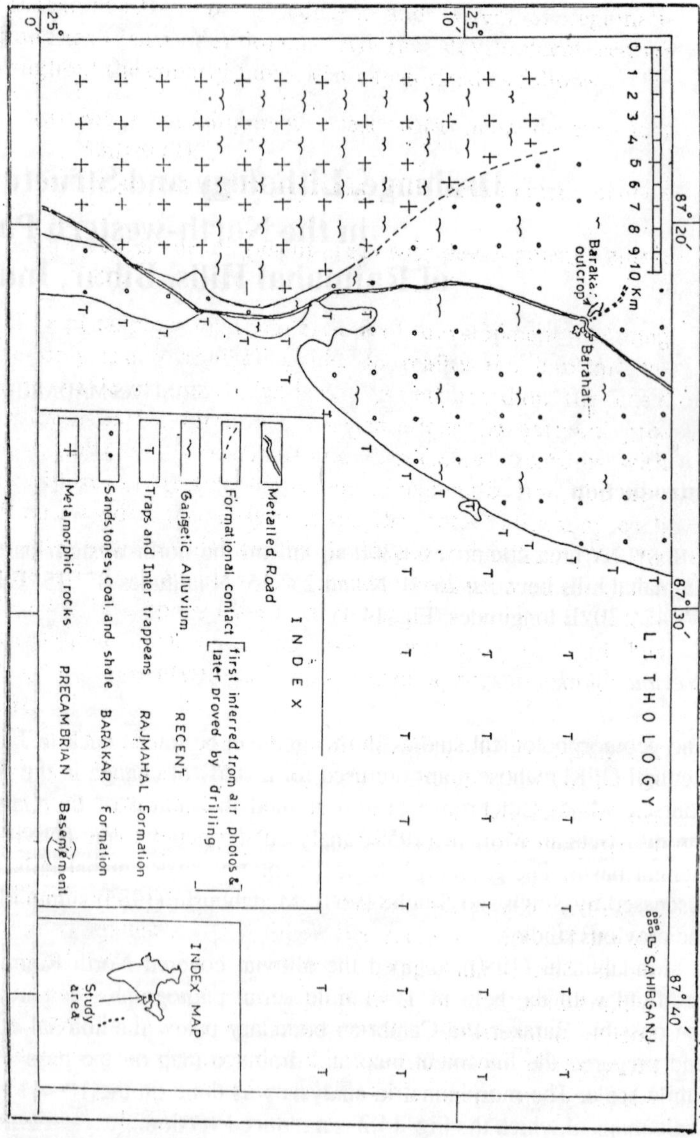

FIG. 14.1: Location of the Study Area.

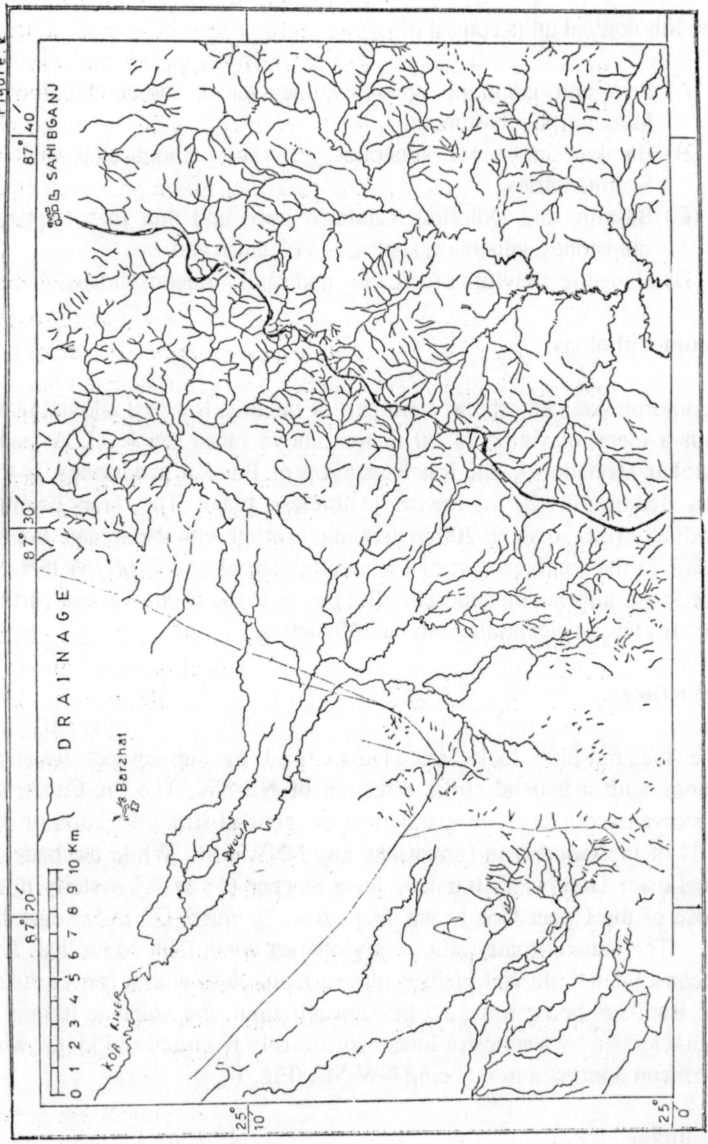

FIG. 14.2: Reduced Version of the Study Area.

Lithology

The lithological units consist of

(A) Pre-Cambrian granite-gneisses; mica-schists and amphibolites – Metamorphic basement;
(B) Barakar sandstones, siltstones, grey and carbonaceous shale – Sedimentaries.
(C) Basaltic and Doleritic Rajmahal traps and thin Inter-trappean sandstones, siltstones, shales; – Volcanics and
(D) Gangetic alluvium of silt, clay and sand – unconsolidated.

Geomorphology

Geomorphologically the western part of the area is a vast alluvial plain with a mean elevation of 50 metres above mean sea level. A small Barakar Monadnock, the lonely outcrop of Barakar sandstones, is the only feature in the otherwise featureless plain. The Trap country gradually rises to about 200 metres in elevation with the arcuate Bejam *pahar* in the southern part of the area having elevation of 261 metres. The Pre-Cambrian metamorphic Bel *pahar* in the south-western part of the area has an elevation of about 200 metres.

Structure

The metamorphic rocks are characterised by sub-vertical foliation planes with a general strike direction of NE-SW. The Pre-Cambrian Gondwana contact is sub-parallel to the general strike direction of the beds of the Gondwana formations viz. NNW-SSE. While the beds of the Lower Gondwana Barakars have steeper (5° to 7°) easterly dips, those of the Upper Gondwana Traps have 'gentler' (3° to 5°) easterly dips. The metamorphic-sedimentary contact is not fault controlled. The unconsolidated alluvial plain is almost featureless with a few faults of the Barakars being reflected through alluvium, the volcanic terrain is characterised by a series of lineaments mostly fractures and joints, with the mean dominant trend being NW-SE. (Fig. 14.3).

Drainage

The river Ganga, lying north of the study area, flows from west to east.

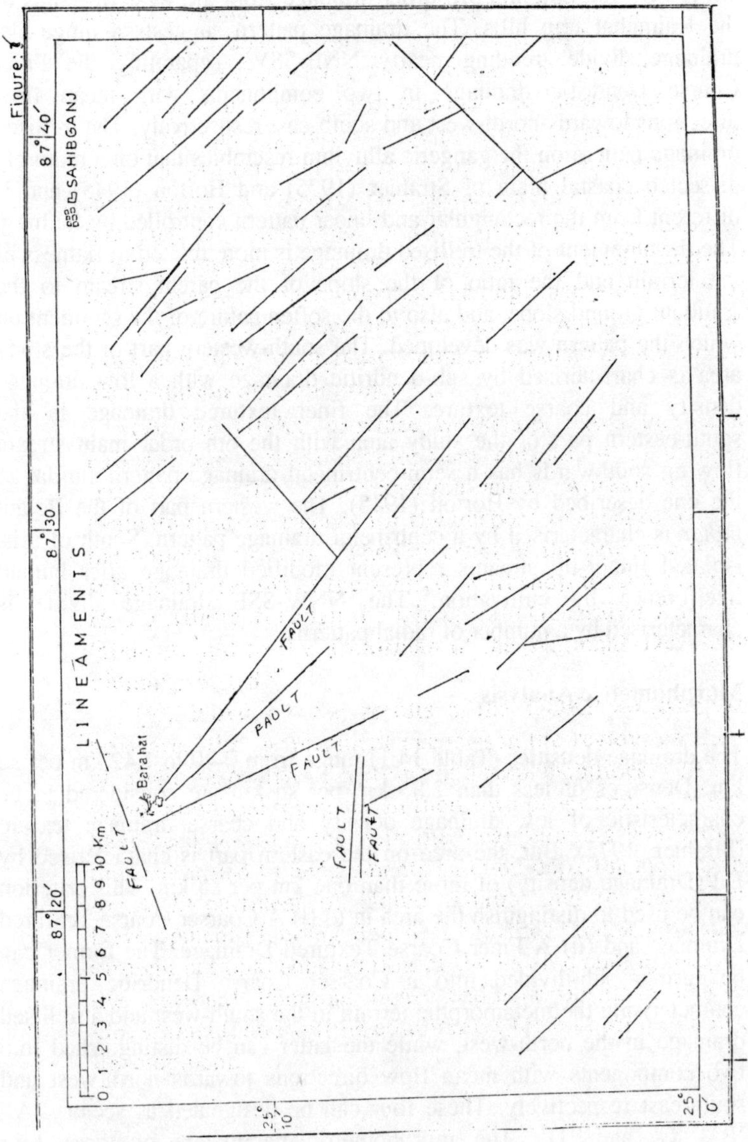

FIG. 14.3: Series of Lineaments.

The westerly flowing Koa and Dhulia drain the Barakar country. These are fed by 1st, 2nd and 3rd order drainage from the western flanks of the Rajmahal trap hills. The drainage pattern suggests a hinge-like drainage divide trending nearly NNE-SSW, separating the Finer Coarse Dendritic drainage in two components with mean flow directions towards north-west and south-east respectively. The trellised drainage pattern on the gangetic alluvium resembles that on a maturely dissected coastal plain of Strahler (1975) and Horton (1945) and is different from the rectangular and linear pattern controlled by faulting. The development of the trellised drainage is more related to flatness of the terrain and the ratio of the slope of the parent stream to the resultant ground slope, and also to the soften nature of the sediment on which the pattern was developed. The south-western part of the study area is characterised by sub-dendritic drainage with a low drainage density and coarse texture. The finer textured drainage in the south-eastern part of the study area with the 6th order main stream flowing southwards has a semi-centripetal drainage pattern similar to the one described by Horton (1945). The western part of the Bejam *pahar* is characterised by a centripetal drainage pattern. South of this, isolated finger-tip streams represent modified drainage after human interference by cultivation. The NNW-SSE drainage divide is characterised by a number of radial patterns.

Morphometric Analysis

The drainage densities (Table 14.1) range from 0-40 to 1.42 km per sq km. Densities of less than 2.85 km per sq km are considered to be characteristic of low drainage density and coarse drainage texture (Strahler, 1975). But, the area on the eastern part is characterised by Dd (Drainage density) of more than one km per sq km. This criterion can be used to distinguish the area in to (i) A Coarser Coarse Textured Drainage and (ii) A Finer Coarse Textured Drainage. The former can be further sub-divided into a Coarser Coarse Dendritic drainage characterising the metamorphic terrain in the south-west and a trellised drainage in the north-west, while the latter can be distinguished into two components with mean flow directions towards north-west and south-east respectively. These four can be designated as sectors 'A', 'B', 'C' and 'D'. The morphometric parameters of these four sub-basins are summarised in Table 14.1. Of the total 1300 sq km of the area, the four sub-basins contribute to 18 per cent, 18 per cent, 29 per cent and 35 per cent respectively. Of the total 1,235 streams recognised in the area, 73 per cent are of the first order, 21 per cent are of second order and 4 per cent are of the third order. The elongation ratio (average 0.75) and Circulatory ratio (average 0.67) are low (Ganesh *et al.*, 1988) and may reflect a possible structural disturbance in the sub basin 'C'.

The Bifurcation ratios average 3 for sub-basin A, 2 for B, 4 for C and 5 for D. The lower value 2 is probably because of the Trellised drainage. The abnormally high (9) value of Bifurcation ratio of the third and fourth order streams of D may indicate a probable structural disturbance (Strahler, 1975).

The semi-log plots of Stream Order, Stream number and average length (Fig. 14.4) on comparison with the plots of Horton (1945) indicate that (a) the Number plots of 'B' resemble Neshaminy, Tohickon and Perkiomen drainage basins of Delware river, Pennsylvania while those of 'A' also resemble the normal plots in a structurally unaffected area. (b) The number plots of C and D differ much from Horton's plots of undisturbed area, suggesting a structural disturbance in the sub-basins 'C' and 'D'. The average length plots also corroborate this.

While sub-basins 'A' and 'B' follow closely the Horton's Law of Stream Numbers (the progression 1, 1/2.26, $1/(3.23)^2$, $1/(2.96)^3$, $1/(2.69)^4$ is closer to the progression 1, 1/(2.84), $1/(2.84)^2$, $1/(2.84)^3$, $1/(2.84)^4$ where 2.84 is the average bifurcation ratio. The deviation from the law is much more in sub-basin D.

Sub-basins A, B and D follow the Horton's law of stream lengths (law of cumulative mean lengths, Strahler, 1975), while in C, the law is deviated much, even for lower order streams. Hence this drainage is not controlled by lithology but by structural disturbance.

Discussion

The Pre-Cambrian terrain of hard metamorphic rocks with uniform resistance is characterised by a 'Coarser Coarse Textured Dendritic pattern', the alluvial 'Marginal plain' consisting of soft unconsolidated sediment is characterised by trellised drainage with low bifurcation ratio. Rajmahal uplands consisting of hard volcanic rocks of uniform harder resistance are characterised by 'Finer Coarser Textured Dendritic Drainage pattern'. While the main controlling factor of drainage in metamorphic and unconsolidated terrain is lithology, structure controls the drainage pattern of volcanic country. The fracture and lineaments of the Rajmahal traps control the drainage.

TABLE 14.1: Drainage Density

Sub basin	Area (km^2	Peri meter (km)	Max. basin length (km)	Circu-latory ratio	Elonga-tion ratio	Stream order (Nu)	No. of seg-ments (Nu)		Nu/N1	Bifur-cation ratio (Rb)	Steam freq-uency (Fs)	Mean length (L_{th})	Cumulative Mean length (L_u)	Leng-th (R_1)	Draina-ge Den-sity Dd
1	2	3	4	5	6	7	8	9	10	11	12	13	14	15	16
A.	239	67	22	0.67	0.79	1	32	63	1		0.35	0.9	$0.9 = 0.9 \times 1$		0.68
						2	23	28	1/(2.26)	2.26		2.6	$3.5 = 0.9 \times 3.9$	2.9	
						3	5	6	$1/(3.23)^2$	4.6		8.1	$11.6 = 0.9 \times (3.6)^2$	3.2	
						4	2	2	$1/(2.96)^3$	2.5		6.0	$17.6 = 0.9 \times (2.7)^3$	0.7	
						5	1	1	$1/(2.69)^4$	2.0		3.5			
						Total	83		Average	2.84		Average		2.27	
B.	234	62	22	0.76	0.79	1	24	63	1		0.16	0.5	$0.5 = 0.5 \times 1$		0.40
						2	6	16	1/(4.0)	4.0		3.0	$3.5 = 0.5 \times 7.0$	6.0	
						3	Nil	0		-		-	$3.5 = 0.5 \times (2.7)^2$	-	
						4	3	8	$1/(2.0)^3$	-		4.6	$8.1 = 0.5 \times (2.5)^3$	-	
						5	2	5	$1/(1.9)^4$	1.5		9.1	$17.2 = 0.5 \times (2.4)^4$	1.98	
						6	2	5	$1/(1.64)^5$	1.0		12.2	$29.4 = 0.5 \times (2.3)^5$	1.34	
						7	1	3	$1/(1.7)^6$	2.0		7.2			
						Total	38		Average	2.125		Average		3.10	

Contd.

TABLE 14.1 *Contd.*

Sub basin	Area (km^2	Peri meter (km)	Max. basin length (km)	Circu- latory ratio	Elonga- tion ratio	Stream order (Nu)	No. of seg- ments (Nu)		Nu/N1	Bifur- cation ratio (Rb)	Steam freq- uency (Fs)	Mean length (L_{th})	Cumulative Mean length (L_u)	Leng- th (R_1)	Draina- ge Den- sity Dd
1	2	3	4	5	6	7	8	9	10	11	12	13	14	15	16
						6	2	5	$1/(1.64)^5$	1.0		12.2	$29.4 = 0.5 \times (2.3)^5$	1.34	
						7	1	3	$1/(1.7)^6$	2.0		7.2			
						Total	38		Average	2.125		Average		3.10	
C.	370	90	33	0.37	0.66	1	441	77	1	4.41	1.54	0.6	$0.6 = 0.6 \times 1$		1.42
						2	100	17	$1/(4.4)$	4.6		1.7	$2.3 = 0.6 \times 3.8$	2.8	
						3	22	4	$1/(4.5)^2$	3.7		3.0	$5.3 = 0.6 \times (2.97)^2$	1.8	
						4	6	1	$1/(4.2)^3$	3.7		6.4	$11.7 = 0.6 \times (2.7)^3$	2.1	
						5	2	1	$1/(3.9)^4$	3.0		0.64	$12.3 = 0.6 \times (2.1)^4$	0.1	
						Total	57		Average	3.9		Average		1.7	

Contd.

TABLE 14.1 *Contd.*

Sub basin	Area (km^2	Peri meter (km)	Max. basin length (km)	Circu-latory ratio	Elonga-tion ratio	Stream order (Nu)	No. of seg-ments (Nu)		Nu/N1	Bifur-cation ratio (Rb)	Steam freq-uency (Fs)	Mean length (L_{th})	Cumulative Mean length (L_u)	Leng-th (R_1)	Draina-ge Den-sity Dd
1	2	3	4	5	6	7	8	9	10	11	12	13	14	15	16
D.	450	91	31	0.68	0.77	1	380	70	1		1.2	0.6	$0.6 = 0.6 \times 1$		1.2
						2	132	24.3	$1/(2.9)$	2.9		1.6		2.7	
						3	27	5	$1/(3.8)^2$	4.9		2.4	$2.2 = 0.6 \times 3.7$		
						4	3	0.5	$1/(5.0)^3$	9.0		6.7	$4.6 = 0.6 \times (2.8)^2$	2.1	
						5	1	0.2	$1/(4.4)^4$	3.0		9.2	$11.3 = 0.6 \times (2.7)^3$	2.3	
						Total	343		Average	4.95			Average	2.4	

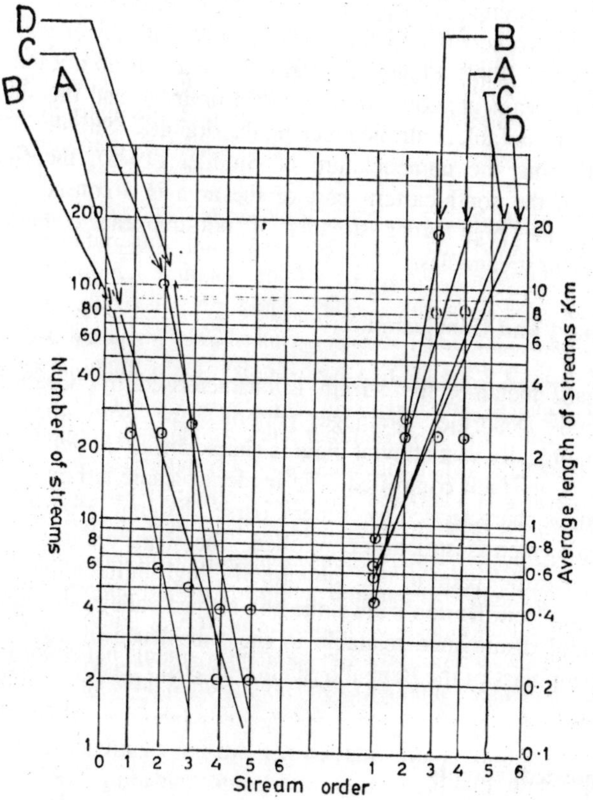

FIG. 14.4: Stream order.

The post-Pleistocene neo-tectonic activity has a clear cut influence on the drainage pattern of Rajmahal traps. The sharp NNE-SSW drainage divide can be attributed only to a structural control. The semi-centripetal drainage of the south-eastern component and the abnormally high bifurcation ratios suggest that the southern part of the trap country was structurally disturbed, probably by the sinking of the great deltaic mass in the south-eastern part of Rajmahal hills (Morgan *et al.*, 1959). The relative uplift of the north-western part of trap country gave rise to the box-shaped valleys and consequent underfit rivers in the alluvial plain. The structural disturbance of south-eastern part of the Rajmahal hills and neo-tectonic activity is inferred by a number of workers (Chowdhury *et al.*, 1989; Qureshy *et al.*, 1989 and Bhave *et al.*, 1989). Photostructural interpretations do not indicate any faults or lineaments controlling trellised drainage and this drainage is different from fault controlled rectangular drainage (Strahler, 1975).

Following the nomenclature of Strahler (1975), the fifth order stream in the south-eastern part of the area is a consequent stream whereas the high order streams of non-trappean country are of subsequent stream class.

Summary and Conclusion

The hard metamorphic terrain is characterised by 'Coarser Coarse Textured Dendritic Drainage pattern', the soft unconsolidated, topographically vast alluvial plain is characterised by trellised drainage which is not fault controlled. The harder volcanic terrain of Rajmahal uplands is characterised by 'Finer Coarse Textured Dendritic Drainage pattern' which is structure-controlled. The sharp drainage divide, the semi-dendritic pattern, the abnormally high bifurcation ratios, the deviation from Horton's laws, the box-shaped valleys are suggestive of structural disturbance relatable to the subsidence of the vast deltaic sediment mass of the Bengal basin during post Pleistocene neo-tectonic activity.

Acknowledgements

I express my deep sense of gratitude to Dr. P.K. Banerji, F.N.A., Dy. Director General, OME & MG Wing, G.S.I., Calcutta, for permitting me to present this paper at the conference, for his constant encouragement and valuable suggestions.

I am grateful to Prof. R. Vaidyanathan, Department of Geography, Andhra University, Visakhapatnam, for going through the manuscript and offering valuable suggestions.

I am thankful to Dr. K.S. Rao, Sri C. Laha, and Sri S.K. Bandyopadhyay, Directors, G.S.I., for their help and valuable suggestions.

REFERENCES

Bhave, K.N., Ganju, J.L. and Ram, Jokhan, (1989): Origin, Nature and geological significance of Lineaments in Qureshy, H.N. and Hinze, W.T., (Edited) *Regional Geophysical Lineaments, Their Tectonic and Economic significance*, Memoirs, Geological Society of India, Vol. 12, pp. 35-42.

Chowdhury, K.C., Das, L.K. and Bose, R.N., (1989): Geophysical lineaments over some geological provinces of India and their tectonic implications in Qureshy, M.N. and Hinze, W.T., (Edited) *op cit.*, pp. 251-262.

Colebrooke, Major, R.H., (1803): On the course of the Ganges through Bengal, *Asiatic Researches*, Vol. 7, pp. 1-32.

Ganesh, A. and Viswanath, V., (1988): Morphometric characteristics of Upper Vaigai Basin, South India, *The Deccan Geographer*, Vol. XXVI, pp. 389-406.

Horton, Robert, E., (1945): Erosional Development of streams and their Drainage basins; Hydrophysical approach to Quantitative Morphology, *Bulletin of the Geological Society of America*, Vol. 56, pp. 275-370.

Madabhushi, Srinivas and Mukherjee, Goutam, (1980): A report on the Geological mapping in the North Extension area of Rajmahal Coalfields, Bhagalpur and Santhal Parganas districts, Bihar, Geological Survey of India, Unpublished Report.

Madabhushi, Srinivas, (1985): Structure, Lithofacies, Sedimentary Tectonics and coal Potentialities of Rajmahal hills, Bihar India, Ph.D. thesis, Andhra University, Waltair.

Morgan, James P., and McIntire, William, G., (1959): Quarternary Geology of the Bengal Basin, East Pakistan and India. *Bulletin of the Geological Society of America*, Vol. 70, pp. 319-342.

Qureshy, M.N., Kumar, S. and Gupta, G.D., (1989): The Himalayan Mega-lineament — Its Geophysical characteristics, in Qureshy, M.N. and Hinze, W.T. *op cit.*, pp. 207-222.

Rennell, Major, James, (1781): The Ganges and Brahmaputra Rivers; Royal Society of London, *Philosophical Transactions*, Vol. LXXI, pp. 91-103.

Singh, R.P. and Singh, R.L., (1960): Geomorphological Evolution of the Highlands of Rajmahal, in the *Geomorphological evolution of Chota Nagpur Highlands*, India, pp. 97-107.

Strahler, Arthur, N., (1975): *Physical Geography*, 4th Edition, Wiley Eastern Private Limited, New Delhi.

15

Slope Instability Problems in Karmi Area of Central Himalaya, India

RAVINDRA K. PANDE

DETERIORATING HIMALAYAN environment, in the form of natural hazards, has become a matter of great concern. Undoubtedly, the structure and still active tectonic process seem to be the prominent causes of slope instability, which, along with the anthropogenic stress, regulate the instability processes. Reports of slope instability in Karmi and its surrounding areas have been reported in past also with noticeable amount of loss of lives and properties. During July 1983 a devastating natural calamity occurred due to cloudburst and heavy rains at Karmi village. The major objectives of the present paper are: (i) to identify present and susceptible areas of instability and (ii) to find out litho-tectonic, geomorphic and anthropogenic causes of slope instability.

The area under study is situated between 29°50′ and 35°0′ N latitude and 79°45′ and 80°0′ E longitude in the Central Himalaya, India. Forty-nine observation sites from seven reference villages have been selected the study (Fig. 15.1). All the seven reference villages have been selected from following seven different lithological belts (Table 15.1).

TABLE 15.1: Reference villages with different Lithological Belts

Lithology	Reference Village	Structure
Joshimath Fm	Dwali	Vakrita Group
Munsiari Fm	Dhakuri	Almora Nappe
Granite-Granodiorite	Khati	Almora Nappe
Nagthat-Berinag Fm	Karmi	Berinag Nappe
Mandhali Fm	Loharkhet	Berinag Nappe
Deoban Fm	Habil	Autochthon Zone
Rautgara Fm	Pathing	Autochthon Zone

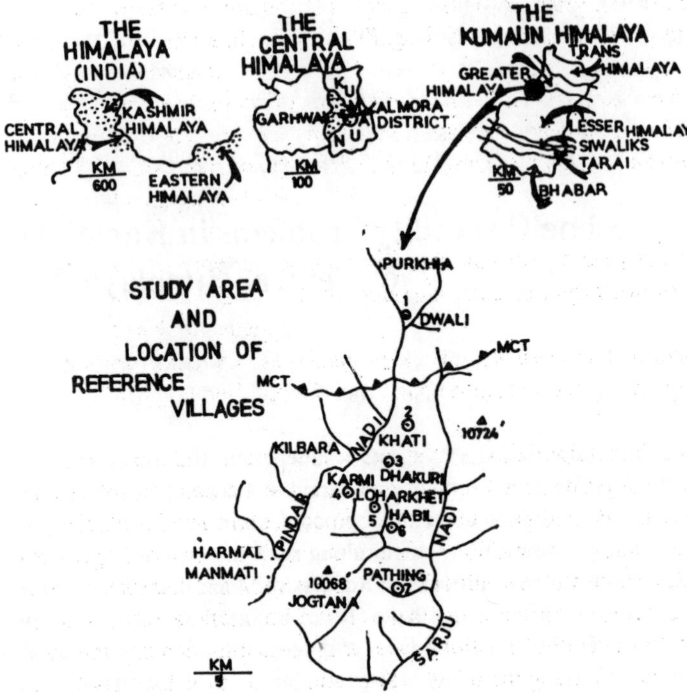

FIG. 15.1: Location of the Reference Villages in the Study Area.

Geology

The rocks in this region belong mainly to Lesser and Greater Himalaya (Fig. 15.2). In Joshimath Fm of Vakrita Group very coarse-grained kyanite and locally sillimanite-bearing garnet-muscovite-biotite psammitic gneiss interbedded with garnetiferous kyanite-muscovite-biotite schists are the major constituents (Valdiya, 1980).

Almora Group has been divided into three units: (a) in the lower part garnetiferous mica-schists, micaceous quartzites and augen gneisses with felspathic schists are the main rocks, (b) the second unit comprises of batholithic and sill-like bodies of sykinematic granodiorite suite intruded by later leucocratic granite and (c) the upper part is consisted of schistose phyllite, carbonaceous or graphitic schists, alternating rhythmically with fine-grained micaceous.

Chlorite schist and amphibolite with inter-beds of thin quartzite of greenish and white colours are commonly found in Loharkhet area under Berinag Fm (Valdiya, 1980). In the north of Loharkhet pink and white or grey, massive, coarse-grained saccaroidal and sparingly sericitic quartzite characterised by lamminae and streaks rich in magnetite are the major constituents of Berinag Fm.

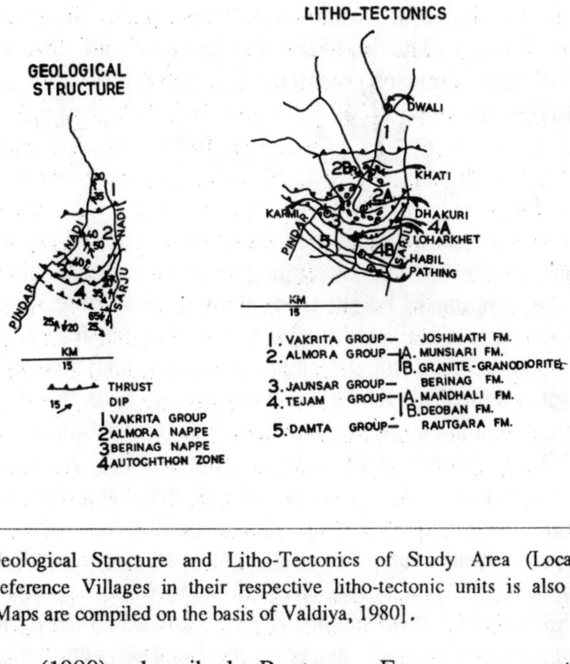

FIG. 15.2: Geological Structure and Litho-Tectonics of Study Area (Location of Reference Villages in their respective litho-tectonic units is also shown) [Maps are compiled on the basis of Valdiya, 1980].

Valdiya (1980) described Rautgara Fm as an extensive lithostratigraphic unit of fine to medium grained muddy quartzite. In Deoban Fm stromatolite-bearing cherty dolomite and dolomitic limestone with grey slate is found in succession. Mandhali Fm is constituted by greyish green and black carbonaceous pyritic phyllites.

Methodology

For the specific purpose of micro-level instrumentation and field observations seven villages from seven different litho-tectonic belts have been selected for reference. After the assignment of code numbers to each reference village, forty-nine observation sites are identified randomly. The numbers of observations are dependent on the intensity, magnitude and nature of instability factors. Border between depressions of streams and open slopes, accumulation of certain types of damage, change in general slope angle, change in landuse and change in slope direction are the main criteria selected for the identification of observation sites. After the assessment of instability in each observation site and investigation of the instability influencing parameters within the observation site types and corresponding degrees of confirmed/inferred instability along with the geographical distribution is registered. Finally, dealing with the types and corresponding degrees of potential damage some management aspects have also been dealt with.

Causes of Instability

It is noticed that on the flanks of the streams, toes of the slopes have developed slip zones almost dipping towards the channels at various angles and ranging in height from 10 to 50 m. These slip zones are most likely to get activated during the subsequent heavy rainfall and would cause sliding of the debris. Though this type of cause is common in almost all the seven litho-tectonic belts but Rautgara Fm and Deoban Fm are more prone to such type of instability.

In certain places debris, filling the valley, is sitting on the incompetent rock surface such as chlorite and talc schists, which are prone to excessive weathering, the instability is further enhanced. Water percolation channelling through these unconfirmable surfaces adds to the mobility of unconsolidated material and such a situation has eventually resulted in subsidence of entire unstable loose rock-mass.

In some cases channels are generally filled up with thick (about 1.5 to 6 m) debris and large size boulders (more than 7 m wide). This accumulation causes daming and diversion during heavy monsoon. In Nagthat-Berinag Fm and Joshimath Fm this factor is found more prominent. The Karmi 'nala', near Karmi village, originally had possibly been narrow (2 to 3 m wide) and deep with high gradient. At present it has, however, been filled up with debris and have become wide (5 to 30 m) and shallow.

In Mandhali Fm and Berinag Fm higher slopes and ridges are comprised of thick sequence of quartzite and slate having acute dip angle (ranging between 40° to 60°). These are few cases where streams are found following the strikes of these formations. Hence, due to scouring in the soft rocks, over hangs of the big chunks of rocks which fall along these slopes after breaking along the major joints causing rock falls all along these slopes.

In the steep slopes and ridges, consisting of thick sequence of chloritic and talcose schists with soapstone and thin quartzite interbands forming dip slopes, sliding and rock fall are quite evident during heavy rains.

The Main Central Thrust (MCT), which is a major structure in the region, separates the Nappe zones from the crystalline Greater Himalaya. According to the available seismotectonic data the Nappe belt of Kumaun Himalaya contains a number of earthquake epicentres. This factor also generally contributes to instability in the area on a regional scale.

Chemical degradation of rocks is also a common phenomenon in this region. In Mandhali Fm pyritous shales suffer oxidation under climatic variation. As a result of it limonite and sulphuric acid are produced. The sulphuric acid converts calcite into gypsum like lubricant material.

Human activity (faulty landuse and deforestation) have grately increased the frequency of instability incidence. Ill-conceived, mismanaged, unrestrained and unplanned human activities have given rise to a large scale mass movement. The cutting of slope for construction have disturbed the angle of repose. In this area, on an average, human activities have increased the incidence of slope instability by a factor varying from 20 to 225.

Apart from these above-mentioned factors abrupt change of slopes, pressure of loose material resting at critical angle, squeezing out or lateral spreading of the soft rocks underlying the larger ones or within the succession of rocks, etc., are some other causes of slope instability.

Types, Corresponding Degree and Coverage on Slope Instability

Seven types of slope instabilities have been recorded in the study area, viz., surfacial erosion (15.41%), rock fall (21.64%), collapsed terraces (25.26%), slide (32.25%), flow (2.21%), slump (2.13%) and complex (1.10%). As none of the seven types of instabilities have more than 35 per cent of area affected, it can be claimed that degradation is still within well approachable reclamation limits. Slide is the prominent type of instability and famous Karmi slide of July 1983 is its prominent example. Observations at the site of Karmi landslide reveals that the main cause of devastation was unprecedented downpour due to cloud burst causing excessive outflow in the catchment. This resulted in rapid scouring of the flanks comprised of debris consisting of soil and rock boulders of all dimensions. This outwash was accumulated in the narrow and deep channel forming a dam of about 10 m height upstream of the lower spur on the right flank of the stream. The blocking of main channel caused a temporary change in the course of water flow and the flooded water gushed over the low spur wiping out everything in its way. With more water accumulation in the channel the above barrier was again burst open with hydrostatic pressure and the stream adjusted to its original channel.

It is also noticed that mismanaged cultivation practices have increased the incidence of collapsed terraces and surfacial erosion. Construction of outward terraces, traditional ploughing methods, unscientific crop-combination, limited choice of crops and use of crude agriculture implements are some of the major factors which help in accelerating process of erosion of top soils (Pande, 1988). Rockfall is the principal type of instability in barren land.

Structurally fragile zones are more prone to instability. In Berinag Fm the sedimentary window rocks of calc-zone have been subjected to various phases of deformation which has resulted in the generation of NW-SE, N/NNE-S/SSW and EW trends of floods. This is a very sensitive zone and almost all type of instabilities are noticed here. The major joint sets developed in rocks (50/75 SE, 130/62 SW, 360/50W, 300/V, 205/40 NE, 300/40) have also developed slope instabilities.

The extensive succession of the stromatolite-bearing, cherty dolomite and dolomitic limestone with bands and intercalcations of blue limestone and grey slate of Deoban Fm is most prone to the slope instability and 42.8 per cent of the total incidence recorded are located on it. However, 26.21 per cent instability incidents are registered in cream white, pink, purple, grey or brown coloured fine to medium grained quartzite of Rautgara Fm. Crystallines of Joshimath Fm, carbonaceous pyritic phyllites or slates with marmorized and plastically folded limestone of Mandhali Fm, chlorite schist and quartzite of Berinag Fm, garnetifarous mica-schists, micaceous quarizites and schistose phyllite of Munsiari Fm and granite-granidiorite of Almora Group have 13.00 per cent, 7.00 per cent, 5.20 per cent, 4.31 per cent and 1.48 per cent of the total recorded instability incidences respectively.

The barren area is characterized by 58.83 per cent of slope instability, while agricultural land and forest areas have registered

30.44 per cent and 10.73 per cent of instability affected areas. Among the forested areas pine forest registers maximum cases of instability (62.42%) while oak and mixed forests have 9.32 and 28.26 per cent of affected areas respectively. Out of the total instability incidence, 42.88 per cent cases are one-way-or-another related to old instability while rest of the 57.12 per cent are reported to be newly developed. The rectilinear slope forms exhibits 56.82 per cent of instability incidences followed by concave (22.06%), steep (9.11%), convex (6.41%) and plan (5.60%) slope forms.

Potential Assessment

Under the potential assessment operation all the observation units are categorised into four groups according to their suspectability with areal coverage, i.e., d_1 in case of future mismanaged landuse the instability is suspected, d_2 up to 25 per cent of affected area is suspected to be unstable, d_3 25 per cent to 50 per cent of affected area is suspected to be unstable and d_4 more than 50 per cent of affected area is suspected to be unstable. There are twelve observation sites where more than 50 per cent affected area is suspected to be endangered. Nine highly sensitive units have 25 to 50 per cent of the affected area suspected to be unstable. In eighteen observation sites up to 25 per cent of affected area is potentially endangered. In remaining ten observation sites degree of instability is suspected to be high in case of future mismanaged landuse or tectonic activity.

Interestingly more than 72 per cent d_4 categories are recorded in the barren land while forest land has maximum frequency of d_1 type of potential instability. In Deoban Fm d_4 category of suspected instability is prominent. However, in granite and granodiorite litho-tectonic belt d_1 category of potential instability is registered. Structurally fragile zones, viz., MCT, joints, faults, fractures and fold axis are more suspected to future instability.

Preventive Measures

On the basis of above analysis following preventive measures are recommended:

* Construction of efficient lined drainage to check water percolation.

* Construction of retaining walls in areas prone to slide.
* Stripping of the left over debris, wire mesh and nets can be used for large boulders which a sliding tendency.
* Afforestation of upper reaches and barren lands possibly with those trees and bushes that have the significant water retaining capacity.
* The exposed toes of slopes should be stabilised by applying soil conservation techniques since most of the slips occur in the loose debris forming the slopes.
* In the landslide zones of increased ground level and porewater pressure it is confirmed that water reduces the shear strength. In such cases, it has been experienced that electro-osmotic stabilisation is quite helpful (Kumar, 1980). The installation of electrodes in suitable locations in respect to the potential slide and application of a voltage potential difference across them will divert the soil water flow from anode to cathode and will drain the debris.
* Good deposits of soapstone are present in the area. Two such deposits near Karmatoli and Majwan villages have already been granted on mining lease to private contractors. It is imperative that systematic mining on the basis of a mining plan duly approved by IBM should only be allowed under strict supervision.

Apart from these above-mentioned measures, specific techniques should be adopted for the specific problem, such as, anchoring of rock, construction of protective nets, plugging or grouting of cracks, with hydrated lime-slurry, cement, sodium silicate solution or bitumen (Valdiya, 1984). It has also been felt that the public policy on instability hazards management should encompass decision-making on the issues: when, where and at what cost the various mix of preventive and ameliorative of palliative measures be applied to mitigate the instability hazards (Swarup and Bose, 1984).

REFERENCES

Kumar, A. (1980): Geological Hazards and related problems of Himalayan Villages. In *Science and Rural Development in Mountains* (eds.) J.S. Singh *et al.*, Gyanodaya Prakashan, Nainital.

Pande, R.K. (1980): Landuse and slope instability in Kafra-Bhaunra Nala of Lesser Himalaya, India. In *Geomorphology and Environment* (eds.) Savindra Singh and

R.C. Tiwari. Allahabad Geographical Society, Allahabad.

Swarup, P. and S.K. Bose (1984) Public policy and legislative measures. Report on Landslides in U.P. Hills, DGM. U.P. Govt., Lucknow.

Valdiya, K.S. (1984): Preventive and remedial measures. Report on Landslides in U.P.Hills, DGM, U.P. Govt., Lucknow.

Valdiya, K.S. (1980): *Geology of Kumaun lesser Himalaya.* Wadia Institute of Himalayan Geology, Dehradun.

APPENDIX I

Observation unit code	Total area observed (sq m)	Types and corresponding degree of instability/suspected instability (%)	Landuse C	F	B
			(affected area in per cent)		
1	2	3	4	5	6
J/Dwali/12/ 0.113	106.1	13.0	22	18	60
		Se 08.4	46	22	32
		Rf 10.2	00	18	82
		Ct 22.6	100	00	00
		Sl 40.2	12	34	54
		Flw 13.7	42	46	12
		S 03.7	62	21	27
		C 01.2	29	15	66
		d_1 10.4	65	12	23
		d_2 32.4	24	22	54
		d_3 49.0	21	21	58
		d_4 08.2	18	06	76
M/Dhakuri/06/ 0.071	84.9	04.3	46	08	46
		Se 13.4	52	20	28
		Rf 32.2	10	22	68
		Ct 10.8	84	10	06
		Sl 08.7	29	40	
		Flw 12.4	46	20	34
		S 16.8	32	48	20
		C 05.7	36	10	54
		d_1 12.8	47	40	13
		d_2 46.4	38	22	40
		d_3 38.6	26	20	54
		d_4 02.2	20	18	62
G/Khati/04/ 0.044	91.7	01.5	22	16	62
		Se 09.4	48	12	40
		Rf 42.7	08	16	76
		Ct 11.0	78	12	10
		Sl 13.4	48	42	10
		Flw 08.6	58	22	20
		S 04.5	46	48	06
		C 10.4	22	32	46
		d_1 38.4	54	32	14
		d_2 22.6	17	46	37
		d_3 20.4	17	32	51
		d_4 18.6	16	22	62

Contd.

APPENDIX I *Contd.*

1	2	3	4	5	6
B/Karmi/08/ 0.091	87.4	05.20	52	16	32
		Se 48.9	48	22	30
		Rf 20.2	17	21	62
		Ct 13.1	78	10	12
		Sl 04.2	04	42	54
		Flw 10.1	61	30	09
		S 02.2	36	22	42
		C 01.3	14	54	32
		d_1 11.4	32	18	50
		d_2 15.8	07	32	61
		d_3 41.6	51	22	27
		d_4 31.2	21	08	71
Mn/Loharkhet/ 10/0.103	97.4	07.0	30	22	48
		Se 16.1	40	30	30
		Rf 11.3	16	08	76
		Ct 04.4	69	13	18
		Sl 08.2	20	54	26
		Flw 11.4	44	34	22
		S 16.2	31	49	20
		C 32.4	23	23	54
		d_1 12.3	39	42	19
		d_2 04.8	54	32	14
		d_3 38.7	21	22	57
		d_4 44.2	11	21	68
D/Habil/40/ 0.446	89.7	42.8	14	32	54
		Se 19.1	42	10	48
		Rf 10.3	30	11	59
		Ct 22.4	52	41	07
		Sl 07.7	40	33	27
		Flw 23.6	44	06	50
		S 08.5	47	31	22
		C 08.4	17	31	52
		d_1 02.4	16	64	20
		d_2 18.1	54	26	20
		d_3 17.2	30	12	58
		d_4 62.3	20	11	69
R/Pathing/20/ 0.232	86.4	26.2	42	08	50
		Se 13.1	45	31	24
		Rf 44.2	20	11	69
		Ct 15.3	68	12	20
		Sl 08.4	29	39	32
		Flw 05.2	41	11	48
		S 07.5	24	46	30
		C 06.3	58	22	20
		d_1 11.5	32	58	10
		d_2 22.8	62	22	16
		d_3 31.5	10	08	72
		d_4 34.2	20	12	68

LEGENDS

1. Observation Unit code — Geological formation/Reference village/ Number of obsesrvation sites/Instability factor.
   - J = Joshimath Fm,
   - M = Munsiari Fm,
   - G = Granite-Granodiorite,
   - B = Berinag Fm,
   - Mn = Mandhali Fm,
   - D = Deoban Fm and
   - R = Rautgara Fm.
2. Types and corresponding degree of instability:
   - Se = Surficial erosion,
   - Rf = Rock fall,
   - Ct = Collapsed terraces,
   - Sl = Slide,
   - Flw = Flow,
   - S = Slumping,
   - C = Complex.
3. Suspected instability:
   - d_1 = In case of future mismanaged landuse/tectonic activity the instability is suspected,
   - d_2 = Up to 25 per cent of affected area is suspected to be unstable,
   - d_3 = 25 to 50 per cent of affected area is suspected to be unstable,
   - d_4 = More than 50 per cent of affected area is suspected to be unstable.
4. Land use:
   - C = Cultivated land,
   - F = Forest land
   - B = Barren land.

16

Slope Suitability and Landuse in Bamanwas Tahsil of Sawai Madhopur District (Rajasthan)

GOPAL LAL GUPTA

Introduction

THE AGRICULTURE landuse pattern indeed is the result of interaction of geomorphic and historic socio-economic factors. In Bamanwas tahsil, the influence of physical features specially slope on agricultural landuse has been investigated and presented, since land utilisation in relation to geomorphology, particularly in semi-arid environment, has received scanty attention of the investigators. However, the valuable work in this direction has been done in Poland.

Slope Suitability and Agriculture

The effect of slope on agricultural landuse may be considered as direct and far reaching. The most obvious direct impact of slope on agriculture is the limitation imposed on cultivation, while the indirect effects of slope are pedological and climatic, including the position of the watertable, development of soil and drainage (Fig. 16.1). The slopes over 2° are usually covered by natural forests. With increasing slopes angle, there is an increase in the slope instability and hence it is essential to understand the relationship between the slope and the landuse category. Keeping in view the extent of suitability of agricultural landuse, the slope analysis and the categories of slope have been using went worthis method (1930) and analysed data has been tabulated in Table 16.1 (Fig. 16.2)

FIG. NO. 16·1.

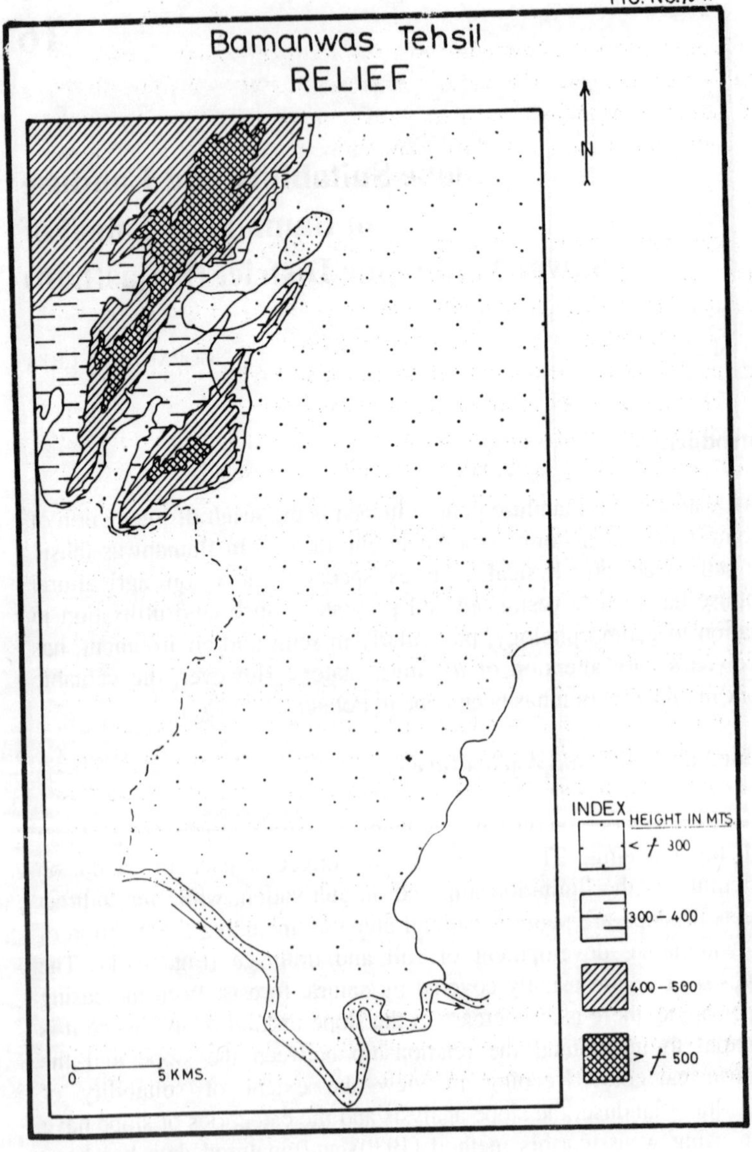

FIG. 16.1: Bamanwas Tahsil: Relief.

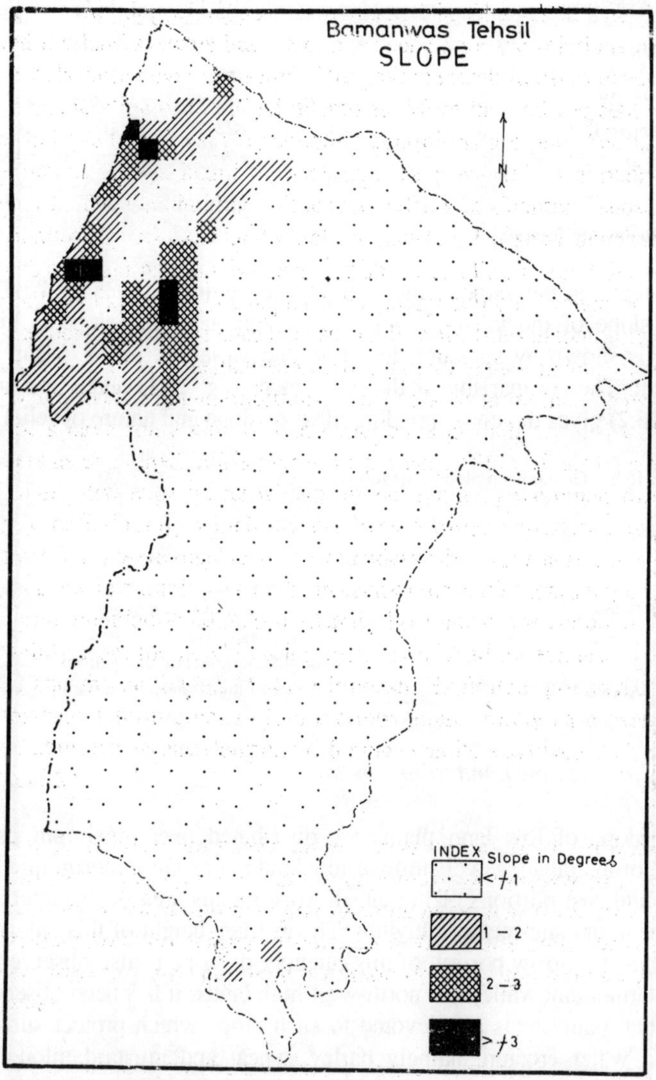

FIG. 16.2: Bamanwas Tahsil: Slope.

TABLE 16.1: Frequency distribution of Slope Value

Sl. No.	Slope Categories	Frequency	Frequency (%)	Cumulative Frequency	Cumulative (%)
1.	Below 1°	743	90.4	743	90.4
2.	1° - 2°	46	5.6	789	96.0
3.	2° - 3°	28	3.4	817	99.4
4.	Above 3°	5	0.6	822	100
		822	100		

(After Wentworth formula)

In the area the slope ranges between a minimum 0.025 to 4°. The mean slope of the tahsil is 0.6°. The variation of the slope is high which is shown by standard deviation and variation, i.e., 476 and 74 per cent. The comparison of the relief map (Fig. 16.1) and slope map (Fig. 16.2) gives the corresponding ideas of slope and nature of relief.

TABLE 16.2: Generalised Slope Categories

Sl. No.	Slope Categories	Level
1.	Below 1°	Level slope of low land plains
2.	1° - 2°	Gentle Slope Zone
3.	2° - 3°	Foot hills of moderate inclination.
4.	Above 4°	Hill tops of moderately steep inclination

Level Slope of Low Land Plains (below 1°)

Level slope of low land plains are distributed over maximum parts (90%) of the tahsil. Level slope of low land plains are in north, middle, south and east portions of the tahsil. Most of this area is characterised by recent and sub-recent alluvium. The average height of this category is 210 m. A narrow portion of this category of slope is also observed in the intermontane valleys of north-west high lands. It has been observed that such plain areas are devoted to such crops which protect surface against water erosion, namely barley, wheat and mustard, ploughed twice a year for seasonal crops. There are some problematic salinity areas.

Gentle Slope Zone (1 - 2°)

A considerable portion of the plain region of the tahsil (5.6%) is covered by this category. It covers the major portion of Singteli, Bhanwra, Bichpari, Raghunathpura and Liwali villages. The gentle slope zone is most productive from the farming point of view under good water supply. These fields are located along the gently sloping valley bottoms and covered with forests.

Foot-hills of Moderate Inclination (2 - 3°)

It covers some part of valley side area of Raghunáthpura, Badh Raghunathpura, Riwali, Amawra, Sikroli Koleta villages and eastern foot-hills of Liwali hills, i.e. Liwali, Kakrals, Raghupura, Jhahra, etc. Villages. The slope zone of 2° - 3° covers and area of 3.5 per cent of the total area. In the extreme southern zone is formed of quartzites but the eastern area is made of gneisses. The general elevation of this category is 400 metres. During the field traverse it was found that the cultivated fields are scattered here and there on lower slopes of ridges.

Hills of Moderately Steep Inclination (3° - 4°)

The highest slope angles (3°-4°) are observed in the north-western part of the Liwali and Bansara villages. In this area it is quite impossible to cut terraces for crop cultivation. Moreover, climatic and soil conditions are too harsh for agronomic practices. Despite this fact some areas of this category have been found under cultivation, confined to valleys and protected places, and which have recently been devoted to the tree crops.

Landuse Mapping Units

In the present case study, the average slope map and an enumeration of aereal distribution of the slope between 1° interval of slope groups indicate that there is uneven distribution of slope throughout the entire area. About 90 per cent area lies in the low level below 1°, occupying 50 per cent of the tahsil net area. While the highest slope ranges from 3° - 4° and covers only 0.6 per cent of the total area.

The landuse mapping units of the study area have been discussed geomorphic region-wise. The area has been divided into two land systems based on the geomorphic characteristics which show the similarities of land forms, land units, surfacial deposits and landuse. The following are the two land systems recognised in the tahsil.

(i) Structural Hills
(ii) Alluvial Plains

(i) *Structural Hill System*

This system consists of hills of the Aravallis and Delhi super groups. The 320 metres contour line marks the lower boundary of the hilly section and the maximum height is about 527 metres at Liwali. These hills are composed of Alwar series. The drainage of the structural hill region is different in many respects. On the whole its pattern is irregular. Most of the rivers of tahsil are structurally controlled. Shallow soils with light texture and yellowish brown colour are found in this region.

The whole system is covered with thickly protected forest. The principal species found here are *Dhok (Anogeissus pendula), Tendu (Deospyros-tomentora), Kher (Acacia catechu) Ber (Zizyphus mauritiana)* etc. These provided the main grazing grounds for the cattle.

Hills and rock plains occupy about 12 per cent of the total tahsil area. These units, due to poor soil-cover and severe water erosion hazard, resulting into rills and gullies are not suitable for cultivation. Certain types of grass and trees, however, come up along the rills and gullies. The inter montane valleys are filled with sediments and gravels brought from hills. So, the villages of Riwali, Amawra, Raghunathpura, Goth etc., occupying the valley are the main agrarian centres of the hilly system. Wheat, maize, rice, sugarcane etc. are the main crops of the system. Barley is important crop which is sown mixed with wheat.

The Alluvial Plain System

This is the largest system of the tahsil, composed of flood plains, natural levels and ravine belt. It is essentially an aggradational plain built of deposits brought predominantly by streams flowing through it. The soils are deep, light textured and yellow brown. The ancient channels are potential sources of fresh water. The watertable is high

(5 m - 15 m) and a number of well water for irrigation as well as for domestic use have been developed.

Badland topography has the separate character of this region. The Morel river and its tributaries such as the Kareli river are characterised by ravine formations. The sysiem has high agricultural potential. *Bajra* is grown in *Kharif* season and wheat during *Rabi* season in these land units. The lands which get flooded during monsoon period, are cultivated only during the *Rabi* season with wheat and mustard crops.

On sub recent deposits, paddy and sugarcane are cultivated in the *Kharif* season. The principal crops in *Kharif* are *bajra*, *jowar*, pulses, and groundnut are cultivated under rainfed conditions and secondary crops such as wheat, barley, sugarcane are grown under irrigated conditions. The area is highly dissected by water erosion, having deep soils with texture varying from sandy loam to loam. Most of the fertile land has gone out of the plough due to fluvial erosion.

REFERENCE

Wentworth, C.K. (1930): A simplified method of determining the average slope of land surface, *Jour. of Sci.* 20.

17

Drainage Density in Relation to Rainfall Intensity in Three River Basins of North India

YASEEN KHAN

Introduction

THE DRAINAGE density is good representative of climatic and physiographic conditions of its drainage basin. It highlights the surface expression of denudational activity within the basin which is influenced by various input and output energies. The drainage density explains the stage of fluvially eroded landscapes. The importance of drainage density in drainage basin analysis is due to two reasons. Firstly it reflects the potential rate at which water can be transmitted through basin, and secondly, it reflects climatic conditions of a particular area and also its basin characteristics including rock types.

In the present paper an attempt is made to analyse the relationship between climatic factor and geomorphic process. The drainage density is considered as a representative of morphometric parameters of land form and mean annual rainfall as climate.

Study Area

The climatic influence on morphometric parameters is analysed in three different river basins under three climatic regimes. For this purpose three river basins — Luni, Mahi and Brahmani of north India – have been chosen, representing the arid, semi-arid and humid climates. These are situated between 70° 30′ to 87° 80′ east longitudes and 20° 35′ to 26° 40′ ' north latitudes respectively.

Database and Methodology

Drainage density-length per unit area is abstracted from third order sample basins on a scale of 1 : 50,000 toposheets. These sample basins are randomly chosen through out the three river basins. The total stream length is calculated with the help of map-reasurer in kilometres and basin area by planimeter in square kilometres. The drainage density is calculated by dividing the cumulative stream length of the basin by total area of the basin. The amount of rainfall of various stations is collected from "Monthly and Annual Normals" of rainfall and Rainy days of Meteorological department. The regression and correlation analysis are carried out between mean annual rainfall, rainfall intensity and drainage density to understand the relationship between them.

Areal Distribution of Drainage Density

The spatial distribution of Drainage Density (Dd) reveals that it varies from place-to-place and basin-to-basin. In the Luni river basin the Dd increases from south-west to north-east directions. The distribution largely is influenced by amount of rainfall and rock types. This distribution shows that high Dd values are concentrated in a area of harder rock. In the Mahi river basin high Dd values are concentrated in the central and north-eastern part of the basin which is comprised by Aravalli and gneiss rock types. The low Dd values are concentrated at the mouth of river basin which is situated on alluvial and blown sand area. The Brahmani river basin also shows the same trend. The drainage density value increases from south to north direction except near the mouth which shows low drainage density.

Areal Distribution of Rainfall

The distributional analysis of mean annual rainfall for three river basins is carried out to understand its pattern. In the Luni river basin, the mean annual rainfall increases from south-west to north-east direction. This pattern shows that the north-eastern hilly region of the basin gets more rainfall than western desertic land. In Mahi river basin mean annual rainfall increases from north-east direction to south-west direction, but interior part of the basin has a lesser value than others. The Brahmani river basin experiences a pattern of increasing mean annual rainfall from north-west to south-east except for a few interior stations, which get more rainfall deviating away from general pattern. This anomaly may be due to physiographic and orographic factors.

The variation of rainfall intensity within the basin is also observed. In Luni river basin rainfall intensity increases from north to south direction. The Mahi basin shows an uneven distribution of rainfall intensity. The high intensity is observed in the periferial area where low values are in the interior part of river basin. In the Brahmani river basin rainfall intensity increases from north-west to south-east direction. Some centrally situated sample basins are an exception to this pattern.

Drainage Density and Rainfall

Climate plays a dominant role in the development of drainage lines of a basin. The degree of development of these drainage lines varies from one climatic condition to other climatic condition and expresses the nature of transformation of landform in a region. The analysis of rainfall within Luni river basin shows a negative relationship with very low correlation coefficient (0.04) (Fig. 17.1A). This relationship indicates that a very low change in Dd takes place over a large change in mean annual rainfall. This is statistically represented by the undergiven equation:

$$Y = - 3.85 - 0.005\,X$$

This contrary relationship may be due to random sampling of river basin, which are representative of the whole river basin. It may be due to peculiar desertic physiographic character of the basin. This above statement is also confirmed by second hypothesis of Stoddart's, who states that "The Dd decreases with increase in rainfall as vegetation density, transpiration, moisture ratio and other factors increase". In this river basin this hypothesis does not apply in terms of vegetation density but in terms of evaporation and moisture retention in the sandy soils.

The rainfall intensity and drainage density show a positive relationship at low level of significance. The increase in the intensity of rainfall is associated with increase in drainage lines because run-off of the basin is directly related to rainfall intensity which is again related to denudation. there by giving higher drainage density value.

This relationship is also confirmed by the following equation:

$$Y = 3.25 + 0.02\,X$$

In Mahi river basin the mean annual rainfall is analysed with Dd to predict the characters of this relationship. Dd and rainfall show a high correlation coefficient (0.54) at 1 per cent level of significance (Fig. 17.1A). This is also depicted by regression equation:

$$Y = 1.97 + 0.002X$$

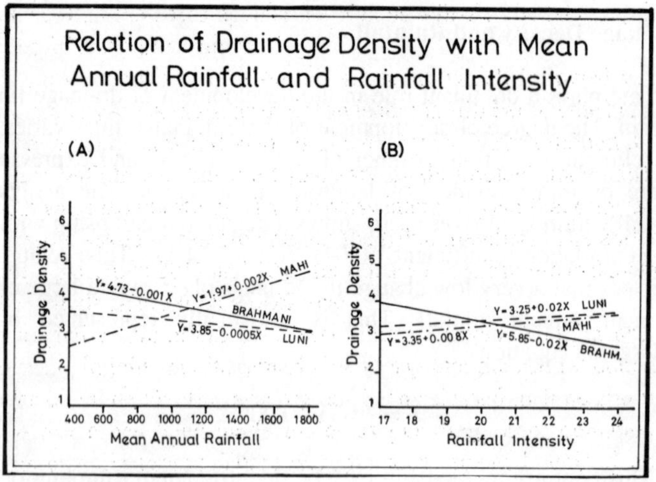

FIG. 17.1: Relation of Drainage Density with Mean Annual Rainfall and Rainfall Intensity.

This relationship reveals that the drainage density increases with mean annual precipitation. Among the three river basins only Mahi river shows a positive relationship with drainage density. This confirms the peculiar nature of the semi-arid climate; supported by other natural factors. Surface wash gives rise to new streams which increase the drainage density in this basin.

Rainfall intensity also shows a positive relationship with drainage density at low value of correlation coefficient (0.02) (Fig. 1B). This negative relationship is due to increase in rainfall intensity which in turn increases the drainage lines but permeability and porosity of rock types and vegetational cover reduce the drainage line variation. This relation is presented by the regression equation:

$$Y = 3.35 + 0.008\,X$$

The humid climate, represented by Brahmani river basin shows a negative relationship of mean annual rainfall with drainage density with low correlation coefficient value (-0.27). (Fig. 17.1A). This is because the humid climate is characterised by high amount of rainfall over evaporation. This leads to increase in the amount of water on earth surface. This also has a simultaneous impact on vegetational growth which increases in direct proportion to the rainfall. This means that increase in mean annual rainfall reduces the drainage density which is shown by this regression equation:

$$Y = 4.73 - 0.001\,X$$

The rainfall intensity shows a negative relationship with drainage density. This relation is depicted by this equation:

$$Y = 5.85 - 0.02\,X$$

This negative relationship occurs due to increase in growth of vegetational cover which reduces the impact of rainfall intensity on surface which in turn reduces the drainage density. The humid climate is only climate which shows negative relationship with both the parameters.

Conclusion

Analysis of mean annual rainfall and rainfall intensity with drainage density in three river basins of north India show varying nature of drainage density. The mean annual rainfall shows a positive relation with drainage density in Mahi river basin (semi-arid) at a high level of significance and remaining two basins show a negative trend. This confirms peculiar nature of semi-arid climate, which upholds the hypothesis that the denudation rate is high and intensive in semi-arid climate but reduces to either side of arid and humid — one due to reduction in annual rainfall and other due to increase in vegetation density. The correlation analysis of drainage density with rainfall intensity shows positive relation in arid and semi-arid climate but negative in humid climate.

Acknowledgements

The author is extremely grateful to his Supervisor Dr. (Mrs) S. Padmaja for her encouragement towards writing this paper and pains to go through manuscript and give valuable suggestion.

REFERENCES

Carlston, C.A., (1966): The effect of climate on drainage density and stream flow. *Internat. Assoc. Sci. Hydrol. Bull.*, 11 pp. 62-69.

Doornkamp, J.C. and C.A.M. King, (1971): *Numerical analysis in Geomorphology.* Arnold, London, 372.

Ghose, B., *et al.* (1967): Quantitative Geomorphology of Drainage basin in the Central Luni basin in Western Rajasthan, *Annals of Geomorphology*, Vol. 11(2), pp. 146-160.

Gregory, K.N., (1975): Drainage Density and Climate. *Annals of Geomorphology*, Vol. 19(3), pp. 287-298.

Khan, Y., (1982): Drainage density as a parameter of denudation Index-A preliminary appraisal of three river basins of north India. *Unpublished M.Phil dissertation*, Jawaharlal Nehru University, New Delhi 110067.

Khan, Y., (1984): The relevance of Drainage Density in Denudational Analysis. *Annals of the Association of Rajasthan Geographers*, Vol. 4, Annual Number, pp. 76-81.

Morgon, R.P.C., (1973): The influence of scale in climatic Geomorphology: A Case study of drainage density in West Malaysia. *Geographiska Annaler*, Vol. 55(1), pp. 107-115.

18

Salient Slope Aspects of a Granite Hill

DEVENDRA KUMAR

A SERIES of small isolated residual hills composed mainly of Chota Nagpur granite gneiss intrusives injected into the Dharwarian strata and exposed due to the removal of sedimentary layer by prolonged interaction of the process of weathering, denudation and mass-movement rises 100-200 m above the undulating surface of the Ranchi plateau and 700-815 m above m.s.l. The Bariatu hill east of Kanke and north of Ranchi is one of them where the intrusive granites are massive with well-developed spheroidal joints including its underside which results from the encroachment of chemical weathering upon cuboid joint blocks (Thomas, 1974) and the microffissures or microcracks. The rocks disintegrate along these potential fissures or joints under the hydrothermal and atmospheric effects thus leading to the formation of 'tors' or dome like hummocks. On the hill slope the massive granite rocks are either covered by regoliths and other weathered materials or frequently form the rocky surface or microfreefaces. The hill slopes are subjected to complex attack by a variety of erosive weapons, water being the most active agent for the removal of waste materials from most of the slopes. The surface of the hill is largely barren except a few occasional patches of grass and few xerophytic species of plants. Mostly the bare granitic rock is at surface and the slope simply reflects the slope of the solid rock beneath as no where the mantle cover is thick enough to form a true talus.

Methodology

The present paper is based on informations received and revealed by four surveyed slope profiles running from the crest of the hill to its base. Each of the profile was surveyed by Pantometer, 1.5m in length using the method outlined by Pitty (1966). The angle of slope and the nature of surface sediment for each unit were recorded and plotted against each other without vertical exaggeration. The obscuring materials of the solid rock elsewhere were classified into clitters, regoliths, mixed debris, rock boulders of varying sizes and sands left behind by slope wash. The sampling and interpretation of the informations enable to assess and estimate the degree of interrelationship between slope form, slope angle and slope type of the Bariatu hill in particular and other granite hills of the Ranchi plateau in general.

Morphological Background

Morphological mapping of the Bariatu hill (based on field measurement and observation) revealed that towards east the hill is in the form of a dome and is surrounded by loose correstones of partially rotted and fractured rocks etched between rock fragments and clitters in the basal section. Elsewhere the bare rock is smooth and is marked by occasional occurrences of tors like groups of weathered boulders. The crystal section of the hill is dominated by alternating appearance of bare rock and rock boulders. Over the junction of bare rock and rock boulders thin veneers of clitter, regolith and sand exist. Thickness of these broken materials depends entirely on the nature of ground slope and the size of obstructing fragments. Towards the south-western upper slope of the hill mixed debris of various sizes are juxtaposed between rock boulders on either side. Convexity of slope in the crystal region is common while concavity is marked on the basal agglomerational section consisting of broken and etched rock fragments, talus blocks, regoliths and clitters and fine fragments of rocks of sands. The hill is surrounded by residual red loamy soil cover. Runoff over the hill is carried through numerous distributary threads developed over its surface following the joints of the rock. Only a few permanent channels have developed which carry rainwater from its mid slope section to the basal section and the contribute runoff to the Jumar Nadi. A conspicuous rock channel has developed over its upper north-eastern slope (Fig. 18.1).

The east facing dome reveals pronounced summit convexity and steep flanking bare rock slope which is the results of circularity of plan (Hack, 1966). Several tors located on different faces of the hill at different heights do not exceed the heights of 10m and conspicuously the tor profiles are highly irregular. Nowhere the tor profile conform to a domical outline. But the little weathered constituent blocks have merged with the true dome (Thomas, 1974) at the western top of the eastern dome.

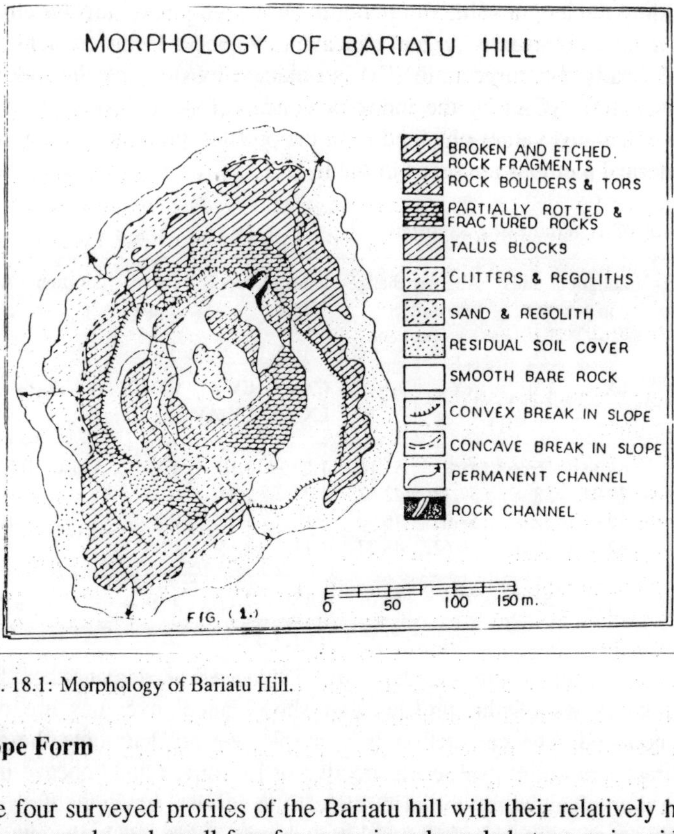

FIG. 18.1: Morphology of Bariatu Hill.

Slope Form

The four surveyed profiles of the Bariatu hill with their relatively high average angle and small free faces are in a degraded stage as is evident from their highly irregular nature (Fig. 18.2). The profiles through their curved and rectilinear units reveals a convexo-concave form. The amount of curvature of profiles is determined by the consideration of the curved segments of the measured section of the profile (Doorncamp & King, 1971). The amount of convexity and concavity of profiles is calculated in percentage by considering the length of convex units (LX) and concave units (LV) with respect to the total length of units of each profile. Since the maximum and minimum segments lying above and below the curved units are included twice, the sum of LX percentage and LV percentage adds up to more than 100 per cent. The total angular change down the profile can be ascertained by summing up the survey discontinuties and percentage of convex (AX%) and concave (AV%) discontinuties and determined with its reference. By dividing the length value of curvature by the amount of angular change the index of convex (CX) and concave (CV) curve forms are achieved. And finally the curve ratio (Cr) is obtained by dividing the index of convex form (CX) by the index of concave form (CV) i.e., CX/CV. The calculated values obtained from the profiles with the application of numerical methods are listed in Table 18.1.

TABLE 18.1: Profile Characteristics

Profi-le No.	Length of units in per cent		Angular change in unit in per cent		Index of curve form		Curve Ratio	Location	As-pect
	LX	LV	AX	AV	CX = LX/AX	CV = LV/AV	CX/CV		
1.	63.33	56.66	48.8	51.18	1.29	1.10	1.17	Bariatu Hill	W
2.	70.94	58.97	57.37	42.62	1.23	1.38	0.93	" "	SW
3.	58.92	52.67	53.09	46.90	1.10	1.12	0.98	" "	NE
4.	62.96	50.00	56.62	43.37	1.11	1.15	0.96	" "	E
Total	64.03	54.57	53.97	46.01	1.18	1.17	1.01		

The values innumerated in Table 18.1 depict the form characteristics of the profiles and show that convexities are more pronounced. The calculation is a simple way to determine the form character of entire profile irrespective of its convex and concave units. The numerical expression also reflects the curving tendency of profiles with lesser degree of profile flattening. Table 18.1 depicts that convexity is relatively larger in all the profiles which indicates the control of resistant surface rocks on the curvature of the profiles. The difference in the convexity and concavity of profiles is not large. The percentage of convex length is relatively large in profile to 2 and 4 where the difference is more than 12 per cent, while in profile 1 and 2 difference in within 8 per cent. The tendency of profile is more apparent in their curve ratio. In this respect all the profiles accept

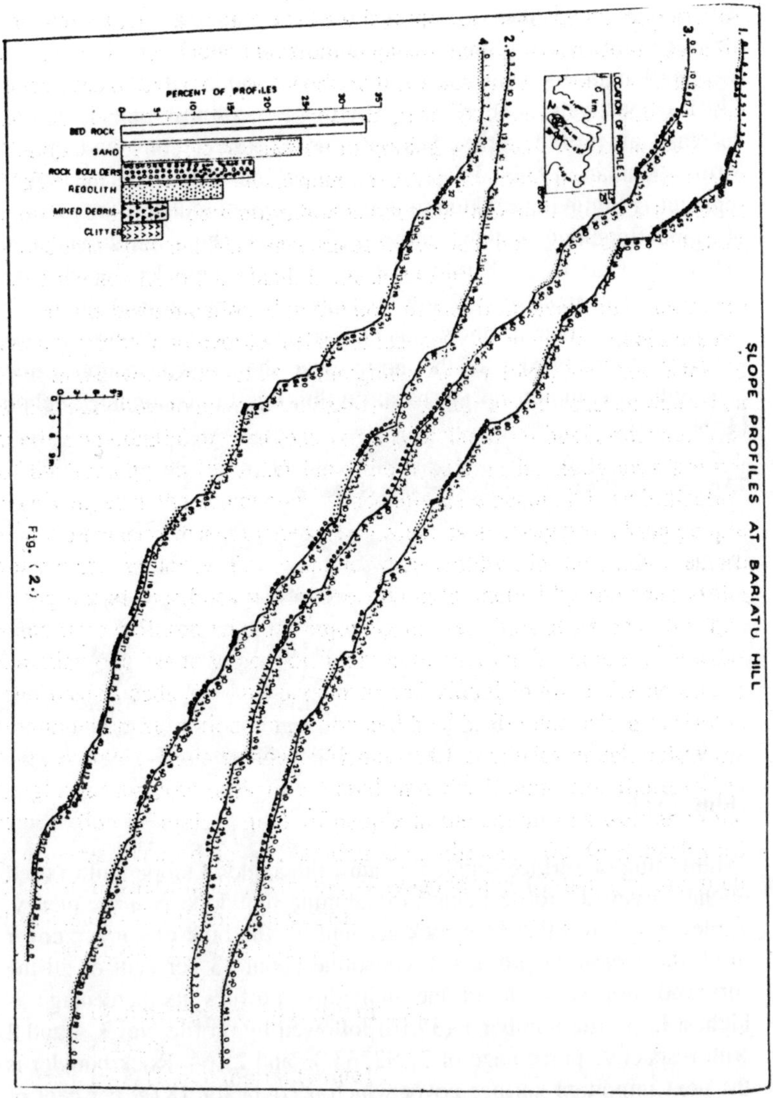

FIG. 18.2: Slope Profiles at Bariatu Hill.

profile indicate concave nature while profile 1 shows the overall appearance of convex form. The percentage of rectilinear or maximum segment is conspicuously low in all the profiles except profile 1 in which the rectilinear segment constitute about 20 per cent of the total segment length. Its percentage in profile 2, 3 and 4 is 16.23, 15.17 and 14.81 respectively. The relative length of rectilinear units tends to decrease as profile approaches towards a perfect convexo-concave forms (Kumar, 1978). In profiles where the length of maximum segment is limited in relation to convex-concave sequence, the process of slope flattening and the advanced stage of slope development is indicated (Carson & Kirkby, 1972). The profiles developed over common structures are not only related to the structure and lithology, but also to the stage of development. The analysis of profile form also indicate the convexity is the norm over the crystal section while concavity is common on the basal portion due to accumulated debris (Table 18.2).

TABLE 18.2: Percentage of slope segment of profiles

Profile No.	Slope segment in per cent		
	Concave	Rectilinear	Convex
1.	40.83	15.00	48.33
2.	52.99	16.23	58.12
3.	55.85	11.71	62.16
4.	49.07	9.20	56.48

Slope Type

The plotting of surface segment against the angle of slope unit in each of the surveyed profile helps to determine the slope type frequency. Table 18.3 shows that bare rock account for the bulk of surface cover in all the measured profiles. It constitute about 33 per cent of all the surveyed slopes, while in the individual profiles its percentage is highest in profile number 1 (39.16) followed by profile No. 3, 4 and 2 with respective percentage of 33.92, 33.33 and 25.64. Rock boulder is the next important surface cover which account for 18.08 per cent of the sampled profiles with the maximum profile No. 3(22.33%). Slopes developed over coarse and fine sands constitute 24.46 per cent of all the profiles with maximum percentage in profile No. 2(30.77%) followed by profile No. 4, 1 and 3. It occupies mainly the basal slopes and the "ledges" where pondage can occur on the upper slopes. Clitters regoliths and mixed debris account for 13.82, 4.39 and 6.22 per cent of the total profiles. Regolith covered surfaces representing past passive areas are now being consumed as the slopes undergo renewed denudation and steepening near individual rises. The limited local relief of less than 150 m permits the back retreat of profiles where minor free faces have evolved due to joints and breaking of the granite rock by physical and chemical weathering. The units of bed rock and rock boulders reveal great variation in slope angles and the maximum slope angles have evolved over them. The debris over the slopes are severely degraded leaving a shallow mantle of clitter (approx. size 5.25 cm) and mixed debris of rocks (approx. size less than 5 cm). Gravity shorting of clitters is locally seen. The percentage of surface cover of the profiles is noted in Table 18.3.

TABLE 18.3: Percentage of Surface cover

Profile No.	Surface cover in per cent					
	Bed rock	Sand	Rock boulder	Clitter	Regolith	Mixed debris
1.	39.16	22.51	18.33	8.33	5.00	6.67
2.	25.64	30.77	19.65	21.37	1.71	0.85
3.	33.92	21.42	22.33	8.93	7.15	6.25
4.	33.33	23.14	12.03	16.67	3.71	11.12
Total	33.01	24.46	18.08	13.82	4.39	6.22

Slope Angle

Analysis of slope angle in the profiles reveal that moderate slope angles are conspicuously more extensive than the gentle and steep slopes. The possible explanation of this fact lies in the character of profile surface and the stage of evolution. The uniclinal and centrally skewed distribution of slope angles in the angle range of 14-25 not only suggest the dominance of moderate slope angles in almost all the profiles, but also depict the lithological control over slope from. The frequency of slope angles developed over different types of surface mantle are noted in Table 18.4 and plotted in Fig. 18.2 which reveal the control of surface mantle over the slope angles of the profiles.

Table 18.4 shows that more than 60 per cent of the slope angles developed over bare rock lie in the category of slope (above 32.0). Its percentage to the moderate and gentle slope is 35 and 5 percent respectively. Rock boulders are also characterised by high frequency of slope angles in the moderate and steep categories, where the percentage of steep, moderate and gentle slope angles are 50, 45 and 5 per cent respectively. Conversely, slope angles of lower categories are dominated by thin cover of sand. In moderate and gentle categories clitters, regoliths and mixed debris are important. More than 75 per cent of the slope angles developed over the sands are below 14.0 and the angles of moderate range constitute the rest, while angles of steep range are rare. Clitter, regolith and mixed debris constitute the bulk of lower and moderate slope angles.

The analysis shows nearly uniform mean angular values between sand mixed debris and regoliths, bare rocks and rock boulders and regoliths and clitters. The interquartile range of bare rock is 16 to above 35, it is 10 to above 36 in the rock boulders. The steepest slopes have developed over the bare rock and rock boulders, gentle slopes over the sands and mixed debris and moderate slopes over clitters and regoliths.

The presence of distinct slope covers with wide variations in slope angles are indicative of the residual hill undergoing active denudation. It confirms the lack of any particular predominant angle of rest for the mixed debris, regoliths and clitters. Rises and minor bluffs over the bare rock and rock boulders mark the joint patterns of the granites where blocks have been separated and dislodged by weathering along the joints. Parallel retreat of slope takes place over these micro-free faces according to the joint and strike planes. These retreating rises do not produce any angle of cohesion thus resulting in the development of shallow mixed debris, regolith and clitter covered bedrock slopes over the hill. Wide variations in slope angles and ground surface irregularities are indicative of the rapid denudation. Sharp fall in angle frequency of 30.0 except bare rock and rock boulders explains that higher angles are produced as a result of rapid basal transportation of accumulated debris. The outcome of denudation can be related to the group of characteristic angles at 22° - 28°. The steeper section of the residual slopes are gradually reduced by denudation to moderate angles of 14° 0 - 26° 0 which are ultimately reduced to relatively gentler angles. Gentler angles are expected to retain their angles for longer period of time (Young, 1961).

TABLE 18.4: Relation of Surface Sediment and Slope Angle Frequency

Surface Cover	Frequency of slope angle in per cent																	
	2°	4°	6°	8°	10°	12°	14°	16°	18°	20°	22°	24°	26°	28°	30°	32°	34°	36°
Bed rock	0	0.66	0	1.32	1.32	0	0.66	1.98	6.62	6.62	0	6.62	0.66	1.98	9.27	1.32	3.31	56.31
Rock boulder	0	0	0	0	2.40	0	3.65	8.43	2.40	9.63	3.61	1.20	3.61	2.40	12.04	3.63	2.42	44.58
Sand	22.53	12.03	9.82	7.25	14.28	7.34	4.57	8.92	3.57	1.78	0.89	3.57	0	1.78	0.89	1.78	0	0
Clitter	0	3.17	1.58	1.65	4.76	4.76	7.93	11.11	4.76	11.11	3.17	6.34	9.52	4.76	9.52	0	3.17	12.69
Regolith	0	0	0	10.0	5.0	10.0	15.0	10.0	5.0	15.0	15.0	5.0	10.0	0	0	0	0	0
Mixed debris	10.71	7.14	10.71	7.14	3.57	7.14	10.71	7.14	3.57	10.71	10.71	3.57	7.18	0	0	0	0	0

The analysis of slope form, slope types and slope angles of the Bariatu hill with the help of slope profiles and numerical methods provide significant clues to the mode of slope development and the association of slope form, angle and surface cover. Several such residual hills with some what uniform structure are found on the Ranchi plateau. The modes and processes revealed in the slope analysis of Bariatu hill would facilitate in the understanding of slopes development in these residual hills.

REFERENCES

Carson, M.A And Kirkby, M.J (1972): *Hill Slope Form and Process*. Cambridge Univ. Press, London.

Doornkamp, J.C. and King, C.A.M. (1971): *Numerical Analysis in Geomorphology: An Introduction*. Edward Arnold London, pp. 136-157.

Hack, J.T. (1966): Circular Patterns and Exfoliation in Crystalline Terrain, Grandfather Mountain Area, North Carolina, *Bull.Geog.Soc.Am.*, 77, pp 975-987.

Kumar, A. (1978): Some Observations of Slope Profiles in the Pat Region, Ranchi, *Symposium Volume on Morphology and Evolution of Landforms*, Deptt of Geog, Univ. Delhi, pp. 41-54.

Pitty, J.F. (1966): *An Introduction to Geomorphology*, Methuen, London, App.A. pp.ii-iv.

Thomas, M.F. (1974): *Tropical Geomorphology*. Macmillan Co. of India Ltd., Delhi, p. 182.

Young, A. (1961): Characteristic and Limiting Slope Angles, *Z. Geomorph* 5: p. 182.

19

The Punjab Piedmont Pediments : A Climato-genetic Interpretation

A.B. MUKERJI

A FAIRLY long and continuous distribution of pediments can be observed along the contact of the south-facing slopes of the Siwalik Hills and the Punjab Piedmont plain to the northwest and southeast of Chandigarh City (Figs. 19.1 and 19.2). The region of their occurrence has had experiences of marked climatic changes and tectonic events and is presently subject to the processes which operate in the Cwg Koppen climatic conditions and occasional neotectonic movements. The origin, evolution, and properties of these pediments would be expected to differ from the classical pediments investigated in the arid and semi-arid regions which have remained tectonically stable over a very long geological period. Although the Punjab pediments are clearly homologous with those occurring along the Aravallis and the Western and Eastern Ghats they are not their analogues. This divergence results from the climatic character of the regions of their occurrence, through climato-genetic processes which have a definite beginning in a particular geological period. Climatic differences, lithologic contexts, tectonic conditions, historical periods of genesis, and topographic associations would distinguish the Punjab pediments from the other members of the genetic group. There is ample evidence of distinctiveness in terms of form, extent, surface appearance, surfacial material, and distribution. Finally, there remains an interesting question: are these pediments being formed presently, are they dead or active?

Significance of Study

Throughout the period they have been investigated, within and outside of India, the pediments occurring in the arid regions have received monopoly considerations. This fact is supported even by a miniscale segment of the continuously bourgeoning geological and geomorphological literature bearing on this feature (Blackwelder, 1931: 442-450; Bradley, 1940: 244-255; Bryan, 1932: 128-129; Bryan, 1935: 765-775; Bryan, 1935-1936: 125-135; Bryan, 1940: 254-268; Denny, 1967: 81-105; Singh, 1977: 6-7; Chatterji, *et. al.*, 1978: 214-215; Singh and Kaith, 1971: 50-59; Kar, Singh and Ghosh, 1977: 16-20; Kar, Singh and Kaith, 1979: 236-245; Kar, 1984: 67-74; Kar, 1984: 21-27; King, 1962: 698-699; and Twidale, 1974: 115-126). A

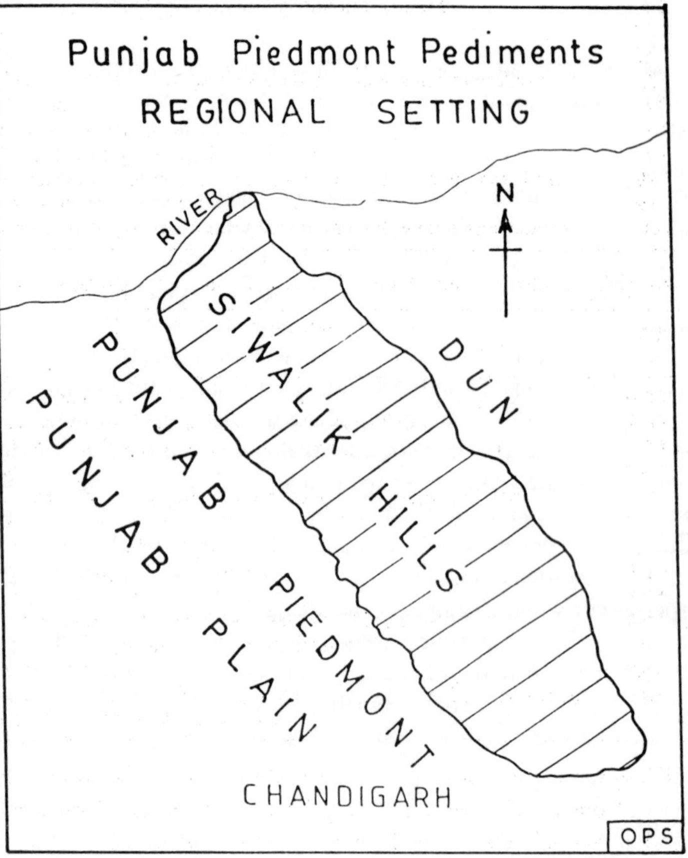

FIG. 19.1: Punjab Piedmont Pediments: Regional Setting.

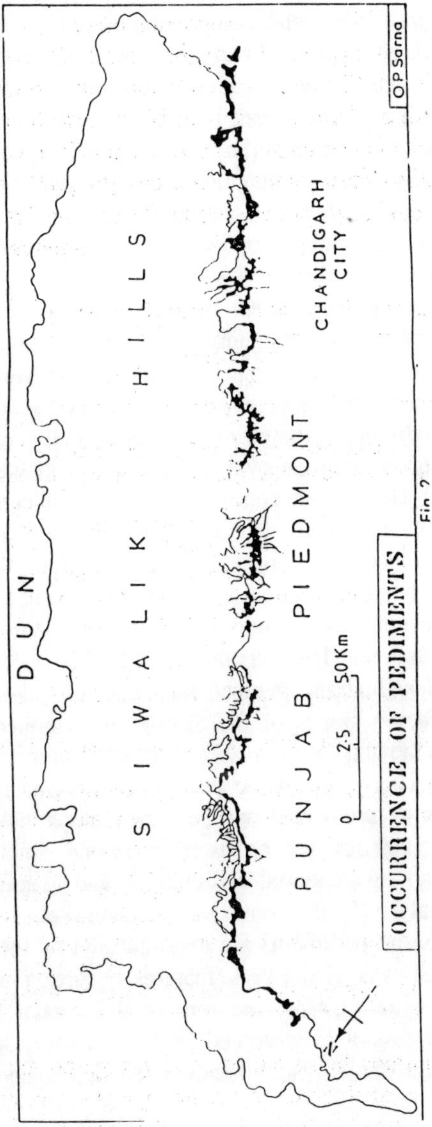

FIG. 19.2: Occurrence of Pediments.

study of the recent literature reveals that the geologists and geomorphologists are slowly becoming aware of the pediment occurrences in humid and cold environments (Dury, 1972: 139-152). It is, however, surprising that the Indian geomorphologists have, by and large, concerned themselves with the arid zone and peninsular pediments. What would attract the Indian geomorphologists to the pediments of the Punjab piedmont is their occurrence in a region that presently experiences Koppen's Cwg climatic type, which is essentially a modification of megathermal, monsoonal regime. These pediments, although not large enough to strike the eyes of the geomorphologists searching for spectacular features, would be interesting for their Cwg climatic associations.

The entire corpus of pediment literature published in India deals with the Rajasthan and the Peninsular Regions which have either experienced phases of tectonic instability in the pre-Tertiary times or have remained stable subsequent to Pre-Cambrian up to the present times. These pediments, therefore, are genetically related to the progress of the pedimentation cycle. In complete contrast the Punjab pediments are developed in a zone that has been subjected to tectonic movements during the Quaternary and Holocene times and have also responded to neotectonic impulses even though the landscape expressions have been subdued. It would be interesting to examine the role, if any, of tectonic movements in the formation and development of the Punjab pediments. This becomes all the more necessary because of the already identified and mapped fault that runs along the foot of the south-west facing slopes of the Siwalik Hills either forming the break-of-slope at the head of the pediment or segmenting its surface.

Almost everywhere the classic pediments reported from different and distant parts of the world are described along with the central Koppice, hill, monadnock, tor, or inselberg around which they form a skirting belt. Both in locational associations and genetic history the pediments are analysed in the context of the central positive relief axis. The Punjab piedmont pediments are distinguished by their location at the foot of the Siwalik Hills, both features occurring in linear belts paralleling each other. Unlike the tors or the central hills and the pediments which cannot be divorced from each other in the genetic sense those of the Punjab piedmont are genetically unrelated to the Siwalik Hills.

Almost nowhere in the pedimont-hill contact ribbon can one find the pediments flanking the *choes* (large, seasonal, ephemeral streams).

Typically, the pediments are laterally disposed along the gullies and extended gullies, both originating on the immediate backing slope or on the dip-slopes hidden behind the southfacing foreslope scarps and terminating at the foot of the scarps or at distances of 150 to 200 metres from it (Fig. 19.2, Plates 19.1 and 19.2).

Finally, the Punjab pediments are made interesting by the absence of convex debris, and free face units of the classic L.C. King's model (King, 1957: 81-102). The absence of thse three essential components of the widely accepted King Model in relation to the pediments clearly illustrated in the field (Plates 19.3, 19.4 and 19.5) will be discussed in the attributes and formation of the pediments.

Location Aspects

For any geomorphic feature, but more particularly that which is surrounded by features of larger dimensions, the two most meaningful locational attributes are site and situation. The Punjab peidmont pediments are much smaller than the peidmont plain in which they are embedded and the Siwalik Hills by which they are both flanked and dominated.

Typically the pediments are situated in the northern rim of the Punjab Plain Piedmont (henceforth termed Punjab piedmont pediment or Punjab pediment) tract that runs in the northwest-southeast direction paralleling the strike of the Upper Siwalik formations which comprise the Siwalik Hills and the direction of the layout of the latter. The tract is dissected intensely by streams of various sizes and types; extended gullies, and gullies of diverse dimensions, and thus displays a rolling appearance. The inter-stream tracts, generally narrow and steep, extend more or less orthogonally to the Siwalik Hills (Fig. 19.2). In the hillward ends of these long, rectangular tracts and at the break-of-slope where the gullies and extended gullies debouch modestly on the piedmont is observed a series of pediments forming a narrow belt running along the foot of the south facing hog-back ridge slopes and scarps.

Along the larger streams which originate deep within the Siwalik Hills and flow through the piedmont and beyond it one encounters *choe* terraces. (Mukerji, 1976: 1-19). The pediments almost nowhere occur along these streams but typically along small gullies and extended gullies, typically consequent, which originate on the south-facing slopes or immediately behind them in the Siwalik Hills.

The streams are typically shallow, about 1 to 1.5 metres deep, about 2 to 2.5 metres wide and 80 to 150 metres long. The gullies are much younger than the streams, probably younger than Holocene, and many of historic ages, about 2000 to 3000 years.

The pediments are sited on the Upper Conglomerate beds, mainly on a bed of clay intermixed with pebbles and cobbles and nodular concretions. The surface below the *in situ* developed grus is hard. (Plates 19.3, 19.6 and 19.7). Pediment sites are characteristically broken and bounded by gullies and extended gullies which are prominently shown in the middle portion as a narrow but perceptible linear depression (Plate 19.11). The distal parts of the pediment sites are terminated at places by a wall-like scarp cut by a transverse gully or by undulated ground of a piedmont (Plates 19.7 and 19.8). The Original site rests on the steeply sloping beds the original configuration of which are still observed at the head of the pediment (Plates 19.3 and 19.5). This relation of the site and the higher bedrock remanants at its back is a crucial clue to the reconstruction of the phases of evolution and the processes of formation. Everywhere the sites are very well drained not only because of the steep dip and slope but also because of the high clay contents of the beds. Where incipient soil formation has started the sub-surface moisture extends up to a depth of only 10 to 15 cm as also indicated by the dark wet colour (Plate 19.6). Very frequently the site is completely dry and bare of incipient soil, grus, and grass cover. It is on these sites that we find the demonstration of the classic pediment-forming processes.

Finally the sites form a long, discontinuous, narrow strip in which the piedmont alluvium at places extends up to the hog-back slopes and at others these slopes drive the alluvium boundary back quite some distance from the Siwalik Hill slopes. The sites thus appear as protruding tongues separated by embayments of alluvium. These can be related mainly to the creation of noticeable terraces by the larger *choes* which just out over a relatively large area from the points of stream debouchement. Thus, the sites of the pediments and the sites of the terraces are locationally completely segregated, suggesting the operation of two different sets of processes at the two different locales.

There is a commanding vista of the piedmont provided from the pediment related to the rather sudden elevation brought about by the steep slopes (Plates 19.1, 19.4, and 19.9). Both pediment strips and piedmont zone are undulating, the rolling character being the product of a large number of streams of various types and size. The numerous fragments of topographic surface are aligned perpendicular to the Siwalik slopes.

Geological Foundation

Though pediment geomorphology, in its character and genesis, has been investigated very largely in terms of the climatic contexts and climatically generated epigenic processes the role of rock types and geological structure providing the initial setting in the evolution of pediment is far too significant to be ignored. It has already been established through research in different climatic regions and in widely separated areas endowed with very contrasting geological contexts and this prompts a critical examination of the climatic contingency versus structural settings (King, 1962: 698-699).

Geologically, the pediments sit on the beds of the Upper Siwalik formation which are comprised to the Tatrot, Pinjor, and Boulder Conglomerate, the first one being the oldest. Most of the Siwalik Hills, above the topographical surface, however, is comprised of different beds and facies of the Pinjor formation (Fig. 19.3). Since most of the pediments have developed on the surface of the Boulder Conglomerate beds it is these which concern us the most. There are almost no sites where the entire thickness of the Boulder Conglomerate is eroded or has disappeared from the surface through fault movements so as to expose the Pinjor and Tatrot beds to the immediate south of the south-flanking slopes of the Siwalik Hills and on which the pediments could have developed.

The Boulder Conglomerate formation, marking the end of the Siwalik deposition during the Middle Pleistocene, forms the uppermost member of the Upper Siwaliks. Stratigraphically the formation sits conformably over the Pinjors (Figs. 19.3 and 19.4). In age it belongs to Lower to Middle Pleistocene (Gansser, 1964: 48). The beds of the Boulder Conglomerate are up to 150 metres thick and are comprised of subangular to subrounded granules and pebble-sized fragments derived from quartzites, sandstones, shales, limestones, charts, slates and schists (Gill, 1985: 45). The fragments of sandstones constitute a higher proportion in the conglomerates of the Upper Siwaliks. Calcareous and ferrugenous nodules occur inside the beds and on the surface.

Essentially the Boulder Conglomerate is comprised of sandstone and clay beds interstratified with conglomerates and often separated by thin pebble or gravel beds. Pink clay and grey and buff sandstones are highly characteristic of the Boulder Conglomerate beds. The beds are consolidated by the infiltration of the siliceous material. There is a preponderance of arenaceous fractions in the lithology (Chaudhri, 1974, pp. 47-51). Hard bed rock appearing as eroded benches with a planar slope are prominent at some places while at other subdued concave slope covered with a thin incipient soil can be observed (Plates 19.2 and 19.5). Whether covered by soil or grus or not ferrugenous matter and clay matrix serve as cementing material everywhere (Gill, 1985, p. 50). Ferrugenous cement forms about one-fifth of the rock composition and hardens the exposed surface to reduced erosion.

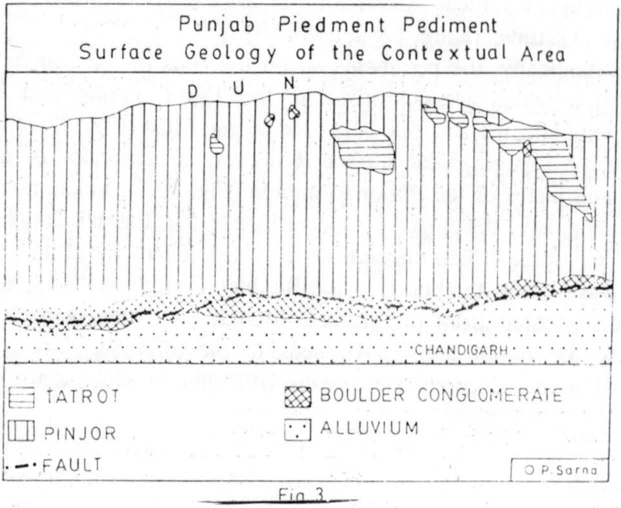

FIG. 19.3: Punjab Piedmont Pediment Surface: Geology of the Contextual Area.

The Siwalik Hills, as a structural form, consists mainly of open, plunging anticlines. Hence, the axial sections of the Hills are gently dipping while the northern and southern flanks are steeper. The Boulder Conglomerate beds, on which the pediments have developed, constitute the southern flanks of the Hills, and extend as a narrow belt hugging the Pinjors which underlie them and the Punjab piedmont alluvium which overlie them (Figs. 19.3 and 19.4).

The Boulder Conglomerate, sufficiently steep to have been eroded into hog-back ridges, dip towards the south and southwest, their dips ranging from lows to about 10° to highs of 40° to 45° (Fig. 19.3 and Plate 19.1). Apart from the open and large anticlines and dips there are several local and regional faults, reversed and normal, which criss-cross the Siwalik Hills. The prominent ones among them extend near or along the peripheries of the Siwalik Hills. The long fault extending along the contact of the Punjab piedmont and the southwest facing slopes of the southern flanks of the Siwalik Hills is indicated by the narrowing of the surface exposures of the Boulder conglomerate beds all along the topographic contacts between the hills and the piedmont, the sharp break in slope, the displacement of beds, and the occurrence of the Boulder Conglomerate outliers at some distance from the head the pediment.

The surfacial material, derived from the bedrocks and created by the epigenic processes, occurrs in various forms on the pediments. Bedrocks outcropping on the pediments is a rare occurrence. In patches one observes weathered mantle rock and at most places coarse young, immature, more endogenetic than exogenetic, soils (Plate 19.2). The presence of soils covered by tufts of grass of long, capillary roots and species of Acacia suggest that the grus and the soil have taken time to develop but not long enough to become mature. After the formation of the pediment which succeeded faulting the terrain must have remained quite stable for a long time, a condition ideally suitable for the perpetuation of the pediments once they have been formed.

Climatic Attributes

The regional climate of the piedmont site and situation is essentially of Cwg Koppen type but its characteristics are a variation, toward lesser rainfall, of the Koppen category. Easily the most significant climatic element involvement in the genesis, evolution, and characteristics of the pediments in the region, as indeed everywhere else, is rainfall, its amount, periodicity of occurrence, fluctuation, and nature.

The average annual rainfall of the region is between 75 and 100 cm but partaking the Cwg rainfall pattern of the north-west India the total annual displays a large range of fluctuation. During 1886 and 1986 the maximum and the minimum of 143.75 cm and 48.75 cm occurred in 1963 and 1883 respectively. This gives a range of 95.00 cm. The large fluctuations in the annual rainfall have been considered as one of the conducive attributes of climate in the genesis of pediments (Beaty, 1974, pp. 19-51; Chatterji, Singh and Qureshi, 1978, pp. 211-224; Singh, 1972, pp. 121-130; and Kar, Singh and Ghose, 1977, pp. 16-20).

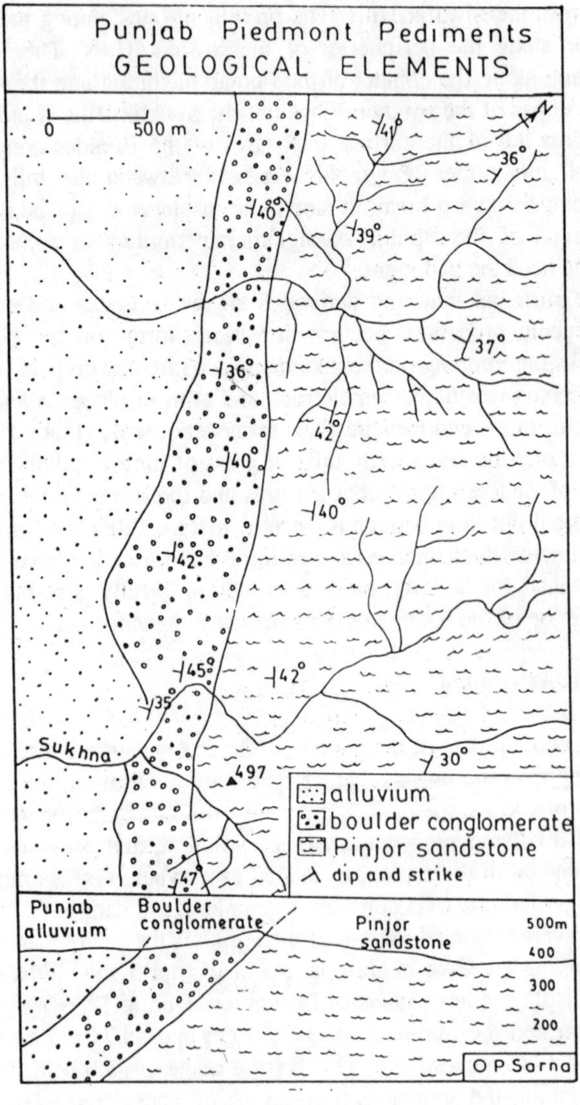

FIG. 19.4: Punjab Piedmont Pediments: Geological Elements.

The fluctuations over the years are equally well reflected in the monthly variations. In any typical year almost 70 to 80 per cent of the annual total falls during the four months of June through September, with maximum of 50 per cent of the annual total always concentrated in June and July. If the sudden, large surface run-off creating the sheet flow is the most crucial determinant of the pediment forming process it is obvious that this process would operate with maximum effect during the months of monsoonal rainfall.

During the last hundred years the trend of the maximum concentration of rainfall during the monsoon months has been both consistent and persistent. Hence, the intensity of rainfall has always been the maximum during the summer monsoonal months. Within these months can also be observed the maximum concentration of the number of rainy days. In a normal climatic year of 40 to 50 rainy days as many as 24 to 27 are found during the summer monsoonal months. In the past the total rainfall in the wettest months has been as much as 175 cm out of the annual total of about 200 cm, thus accounting for 87 per cent. The largest rainfall of 24 hours also has occurred in these months. The intensity of rainfall (I_R = (Rainfall/Days) x 100) in these months can rise to as high as 75 to 80 per cent.

It is during the summer monsoon months that one observes high intensity rainfall caused by thunderstorms, intense local convection, and thermodynamic advective-convective-orographic rainfall. Enormous runoff forming sheetflows are the widespread results.

With the high intensity and high concentration of rainfall during the monsoon months the pediment forming processes attain their maximum power and effect. In the remaining months *in situ* weathering is more prominent, the role of which is only minimal and preparatory.

It is important to remember that the present climate is very different from that of the past. Immediately following the deposition of the Boulder Conglomerate and after the uplift of the Upper Siwalik beds the climate, for a long time, remained cold and dry. This condition persisted through the Mindel-Riss Interglacial. This cold and dry phase is indicated by a complete absence of fossils, red pigmentation, kaolinisation, and calcarious contents of the sediments. (Gaur and Chopra, 1984: 353-355). The climate ameliorated substantially during the Riss-Wurm Interglacial and the later glacial phases. However, the drier periods continued during the Upper Pleistocene (Zeuner, 1953: 242-253). The typical monsoon established itself well after the disappearance of the low-level mountain glaciers, deep within the Holocene.

Dimensional Attributes

The pediments occur in a narrow zone suggesting their relatively small size. Within this zone they do not form a continuous tract but appear as fragmented occurrences (Fig. 19.2 and Plate 19.12). The fragmented nature is not only attributable to the traversing gullies and extended gullies but also to the tongues of exposures of the bedrocks, mainly the boulder units of the Boulder Conglomerate. At the points of debouchement of the major *choes* the pediments tend to disappear and be replaced by terraces (Mukerji, 1976, pp. 1-19). The size and shape of the pediments and their extension very large depend upon the alignment of the gullies and the extended gullies and the appearance of the underlying Boulder Conglomerate beds and their dips.

The size, length, and width of the pediments vary considerably. In length the variations are from as low as 10 metres to the high of 150 metres or even more. Since the fault that traverses the Boulder Conglomerate has a sinuous alignment the length of the exposed bedrock varies considerably along with that of the pediments. Where the fault runs through at some distance from the break-of-slope a longer trajectory plane of the bedrock has outcropped and a longer pediment has developed (Plates 19.2, 19.7 and 19.10). It is suggested that the alignment of the downthrown blocks of the south of the fault has also resulted in the decline of their dips and hence increase in the length of particular slope segments developed on them. At some sites this has not happened and the exposed bedrock reveals a much shorter length. In the field one is struck by the repetitive occurrence of discordant relationship between the length of the exposed bedrock, the length of the pediment, and the transformed dip as a result of the changes in the disposition of the downthrown southern blocks along the fault. Nonetheless the length of the exposed rock and the length of the pediment seem to bear a harmonious relationship at most sites. In the field one also finds an interesting relationship between the direction of the hill-slope behind the pediment and the direction of the *choe* flow or gully flow with the length of the pediment. The gullies flow to the southwest, south, and southeast. The general slope of the Siwalik Hills southern flanks is toward the southwest. Hence the pediments associated with the gullies flowing to the southwest are longer as the gullies in conjunction with the topographical slopes flow over a longer distance. In contrast the gullies which flow toward the south and southeast are discordant with the direction of the topographical slope and are shorter and hence the pediments associated with them are shorter (Fig. 19.5). Thus, the pediments extending orthogonal to the grain of the Siwalik Hills are the longest and those emerging out at lesser angles are the shortest. The more acute is the angle of spread the shorter is the length. This relationship is essentially dependent on the physics of the river flow. The lower angle of emergence necessitates a longer path of the stream on the hill-slopes flowing over which the gully loses a large part of energy through friction and hence its eroding power down the slope and down the pediment zone is reduced. This reduction results in the shortening of the gully and the pediment alike.

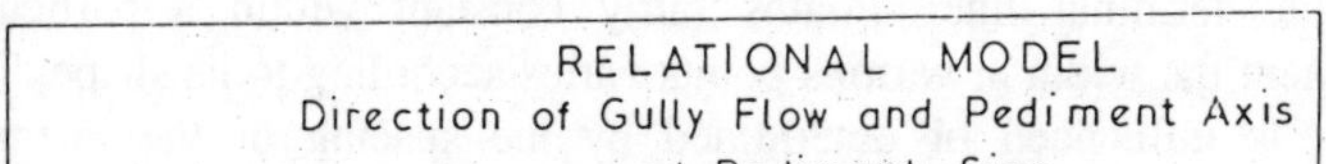

FIG. 19.5: Relational Model: Direction of Gully Flow and Pediment Axis and Pediment Size .

Unlike the length of the pediment which is controlled by both the nature of the bedrock exposure and direction of gully flow the width is determined by the alignment and spacing of the gullies. The field observations clearly reveal that the pediments are wider near their head, the piedmont-hillslope contact, than in their distal parts. The field observations suggest that the pediments generally are sited between the gullies, either convergent and confluent or parallel to each other below the distal part of the pediment. Where the confluence takes place very near the piedmont-hillslope contact the extension of the pediment below the confluence is almost entirely terminated at this point. The distance of the confluence of the adjacent, mainly first-order gullies is a crucial determinant of the variation of the width of the pediments. By the nature of the alignment of the gullies the width of the pediment varies in different parts, proximal, middle and distal. In almost all the pediments the length is indicated by the major axis, more or less, extending through the central parts. Where the pediments are trapezoid in form the length is the same in all directions.

The width and length relationship of the pediments appears to be inconsistent. In some of the large pediments both length and width are large while in others the pediment axis is short but their lateral extension is pronounced. While the length of the pediment from its head to terminal line remains fairly constant within a particular pediment the width at various points varies according to its shape. The width is influenced or determined by the spacing of the laterally flanking gullies, alignment of the bedrock exposures, the channel form of the streams (whether stable or shifting and straight, sinuous, or meandering), stream pattern, pattern of the angle of confluence and projections of the piedmont alluvium. By far the most important determinants are the spacing and pattern of the flank gullies. This simple relationship is repeatedly observed because the gullies hug closely the lateral edges of the pediments to which they are bound genetically. The pediments of the study area vary between less than 100 metres to more than 400 metres. Most pediments being ellipsoidal, semi-circular, and triangular have width of about 100 to 250 metres or so. In the strikingly rectangular pediments the width can range from 250 to 600 metres but these pediments are few in our region. It cannot always be observed presently but it can be argued that the lateral extension was effected by the lateral sweep of the overbank stream flow. In a miniature form it is observed in the region during the few hours or even less than an hour when the enormous discharge collected in the headward slopes as a result of thundershowers seeks to flow out through the rather inadequate channels of the gullies and extended gullies. Splays then laterally sweep the bedrock surfaces.

There is a large diversity of shapes of the pediments that can be unmistakably observed in the field and in the aerial photographs (scale of 1 : 30,000) and also inferentially on the toposheets (scale of 1 : 63,360 and 1 : 50,000). Some of the commonly observed shapes are radial (Plates 19.1 and 19.2), triangular with base along the slope-foot contact (Plates 19.5 and 19.9), trapezoid (Plate 19.7), spindle, semi-circular, narrow rectangular, and ellipsoidal (Plate 19.12). It is very difficult, if not altogether impossible, to generalise the shape geometry, quantitatively measure the shapes, and determine the explanatory variables controlling the shapes. However, it can be conjectured, although it does not appear to be obvious, that the alignment of the bedrock exposures, direction of flow of the gullies, ground plan of the gullies, geometrical design of the channels (straight, meandering, or sinuous), confluent or parallel stream patterns, and the angle of confluence of the confluent streams. All the determinants can be identified although their operation in the determining of shape cannot be fully observed or established in the field. Some are however quite clear. Trapezoid pediments occur between the parallel streams, a triangular with a base at the slope foot is found between two gullies having a low angle confluence and a semi-circular and an ellipsoidal occur between curvilinear alignments of confluent streams. The extension of the Boulder Conglomerate bedrock beyond the south-facing hillslopes of the Siwalik Hills is another important determinant of the pediment shape, the extension itself is contingent upon the initial erosion by adjacent gullies and extended gullies. It is in the interplay of random behaviour of the determinants that the explanations of the shape are located. The randomness is very strikingly demonstrated in a variety of weird shapes of the pediments (Fig. 19.2). At many places it is indeed difficult to distinguish the individual pediments and separate them one from the other. The jumbled combinations reveal a bewildering variety of shapes mainly related to the erosional configuration of the bedrock exposures and the complex pattern of stream and gully integration. At places the complexity of shapes is further complicated by the hill-ward projections of piedmont alluvium. At places over considerable length the bedrocks have been very sharply cut and displaced vertically by a fault (Plate 19.12). At these sites the pediments are invariably broad-based along the slope foot and backed by triangular faceted ridges – a direct evidence of a strong, linear fault. Structure here has played a role in determining the pediment site and shape. This finding in the field and in the aerial photographs fully supports the earlier findings and generalisations (Twidale, 1967: 393-411). Through its relationships with bedrock exposures and gullies the pediment shapes have some genetic implications as well.

Slope is easily the most important element in the morphological personality of pediment as it indeed is in the locationally and genetically associated alluvial fan. Lithology, surfacial material, weathering, nature of run-off, climatic effects, and tectonics are associated with slopes of the pediments and alluvial fans. Geologists have often considered the pediments and alluvial fans either to contrast or to complement their property analyses (Denny, 1967: 81-105). Pediment, equally significantly, is one of the four integral elements of the classical hillside slopes and its evolution and extension are related to the retreat of the scarps into the hills and finally their reduction. It is clear therefore that the evolution and life history of a pediment has to be investigated along with the changes in the other three elements of he slope system.

The pediment surface morphology does reveal three parts, distinguishable in many pediments, proximal, middle, and distal. In most of the members the first and the third are pronounced and easily identifiable because of their location, gradient, and surfacial material.

The proximal segment of the pediments stretches from the hill-slope foot along the contact of the Punjab piedmont and the Siwalik Hills. It is characterised by slopes of 7° to 10° and at many places of up to 18° to 25°. At most places the proximal segment appears to have a subdued concave form but planar forms are also common (Plates 19.2, 19.3, 19.4, 19.5 and 19.10). The bedrock scarp standing immediately behind the head of the pediment has a sharp junction with it at a fairly high angle of 135° to 152° (Plates 19.2 and 19.3). Indeed the presence of this scarp and the high angle of junction immediately signify the presence of the pediment and the origin of its proximal segment. Typically the pediments have an inverted triangle slope which means therefore that we have a wide proximal segment but there are innumerable examples of radial plan proximal segments. Particularly where an extended gully jumps from the hillside slope on to the pediment and is fragmented into distributary gullies the radial plan of the proximal segment becomes strikingly pronounced (Plates 19.2 and 19.7). In those pediments, and there are many examples of them, where scarp retreat is continuing albeit at a slow rate the proximal segment is live and active and is covered with a thin clay-dominant grus that does not remain lodged there for long. At other sites where the scarp is either stable or has become gentle the production and accumulation of grus brought down as colluvium are directly reduced and the proximal segment is practically bare. There is a definite relationship, although not quantitatively determined, between the debris size with the angle of slope on the one hand and distance from the head on the other (Kar, *et al.*, 1979: 236-245). On the proximal segment the surfacial material is fine clay and silt as here the angle of slope is the largest (also clay has a very large angle of repose), the locus is near the head, and the adjacent bedrock outcrop is the clay bed in the Boulder Conglomerate Formation. Unlike the Rajasthan - Aravalli pediments their analogues of the Punjab piedmont are not separated from the hillside slope by talus slope (Kar, 1984: 67-74). The absence of the talus slope revealing an abrupt junction of the pediment with scarp has also been noted by earlier workers in their investigations of the pediments in the tropical savannas which are quite similar to the piedmont characteristics of the study area (Kesel, 1973: 97-109). It is equally true that the proximal pediment represents more or less the original surface as contrasted with a surface of bedrock that has been exhumed from beneath a thick alluvial or colluvial cover (Cooke and Mason, 1973: 49-61).

In total contrast to the proximal segment can be observed the distal segment of the pediment. The distal segment, as the term indicates, is the lower part of the pediment beyond which stretches the alluvial piedmont. Although all the three pediment segments are slopes of transportation, in some measure or other, the distal segment displays a subtle positive balance in favour of debris supply as compared to debris removal. In these distal pediments can be observed a debris cover about 10 to 35 cm thick mainly comprised of coarse sand, pebble, gravels and even small boulders, 6 to 10 cm long (Plate 19.6). These clastic sediments have been partly derived from the proximal segment and the scarp slope behind them and partly from the *in situ* developed grus. The distal pediment has a slope of less than 1° to 1.5°. This gentle slope matches well with the natural angle of slope of the large clastics which, as a result, come to stay for long, until at least the time when debris removal is much faster than debris supply. The distal segment, in itself, is gently planar but joins the middle segment at an extremely low angle, at most sites almost imperceptible.

Between the higher and steeper proximal segment and the lower and gentler distal segment lies the fairly wide middle segment. This segment has a slope of 2° to 4°, at most sites the values hover around 2° or a little higher. The form of the segment is basically planar but toward the upper and lower extremities can be observed subdued minor curves which mark the continuous stretch from the proximal to distal segments. It is in the middle segment that one observes a mixture of finer-textured and coarse-textured clastic debris, the former descending from the proximal segment and the latter contributed both by higher slopes and *in situ* disintegration of bedrock. This is truly the segment of transportation. The annual sheetflood and slow soil creep and mass-movements along with minor gullies transport the sediments down the slope. It may be noted that the fall in elevation from the head to the toe is almost 5 to 8 metres.

It bears repeating that the Punjab piedmont pediment is a part of a truncated classical slope system in which the free-face and waxing slope are totally missing. In the field one is struck by the repetitive occurrence of the steep, constant slope or slope of transportation and the pediment stretching below it. The constant slope has a remarkably concordant relationship with the steeply dipping beds (Plates 19.3, 19.4 and 19.5). In contrast the pediment with a much less steep slope has a discordant relationship with bedrocks. At many sites the constant slope sits on the scarp slope of the hogback ridge and is thus an anaclinal slope, the structural dip matching with the topographic slope. The proximal, middle, and distal segments in combination produce the well-known pediment form of a concave profile, the upper one far steeper than the lowest one. There is a slight rise in ground on the distal parts, but on the whole the concavity persists.

Genetic Interpretation

Studies on the formation of pediments have a respectable age in geomorphology. One of the earliest investigations was attempted in 1897 (McGee, 1897: 87-122). The studies are being continued unabated as investigations are spanning a large diversity of lithological, tectonic, structural, and climatic types. It is largely because of the diversity of contexts that such a large number of hypotheses or theories of origin have been proposed. Of the three climatic-pedologic contexts, arid and semi-arid, tropical savannah, and humid, and first has claimed 32 theories, while the second and third have spawned 20 and 18 theories respectively. It was only in 1938 that the humid (Davis, 1938: 1337-1416) and in 1948 that the tropical savannah (Fair, 1948: 71-79) attracted the attention of the geomorphologists interested in the problem of pediment formation.

Essentially, the theories can be classified into five groups: sheet flow, backweathering, lateral planation, exhumation and composite action. Although a particular group is more effective for a particular climatic context, in most cases it is a combination of more than one theory rather than a single theory that provides the most convincing explanation on both logical-deductive and empirical grounds. The sheet flow theory first propounded by McGee had continued to interest the geomorphologists till the early 1980s, its life spanning almost a century. The interesting fact, however, is the application of all the groups of theories of more than one or even all the climatic contexts, leading to the suggestion that the processes proposed by them are climatic analogues of each other.

In order to identify the theory or the combination of theories which will explain best, at least tentatively, both the formation and the attributes of the Punjab piedmont pediments it is important, at the outset, to determine the approximate geological period of their formation. Three field observations suggest that the pediments are Upper Pleistocene and Holocene and more likely late Upper Pleistocene and Holocene in age: (a) the dipping Siwalik beds were tilted only in the Middle Pleistocene (Mukerji, 1979: 98-107; and Mohapatra, 1979: 107-135), (b) the pediments are everywhere observed flanking the extended gullies and *choes* which originate much south of the main water-divide in the Siwalik Hills and are consequent upon the youngest south-facing hogbacks, and (c) the pediments have everywhere developed on the Boulder Conglomeratic beds.

It was only after the Boulder Conglomerate beds were uplifted and tilted during the later phases of the Middle Pleistocene corresponding roughly to the Mindel Glacial and Mindel-Riss interglacial that the climatic and geological context of the pediment genesis could be established (Fig. 19.6). It should be noted, however, that the Riss and Wurm glacials and Riss-Wurm Interglacial were less humid even though the surface run-off and rainfall were not inconsiderable. (Mohapatra, 1979: 110-111; Mukerji, 1976: 12-13; Vishnu-Mittre, 1965: 332; and Zeuner, 1953: 242-253). These climatic fluctuations have been worked out by several investigators (Mukerji, 1976: 13-14 and Mohapatra, 1979: 110-111).

There were two tectonic events having crucial bearing on the pediment formation: (a) the hog-back ridges on which the pediments sit as concordant with anaclinal slopes were created after the structural deformation of the Boulder Conglomerate took place during the Mindel Glacial and Mindel-Riss Interglacial; and (b) there was a fairly long, sub-regional faulting within the Boulder Conglomerate beds to the south of the main mass of the Pinjor formation (Fig. 19.3), throwing a narrow block of the Boulder Conglomerate bed down toward the piedmont plain. The fault must be one of the events of the neotectonic movements recorded in the vicinity of the southern flanks of the Himalaya (Nakata, 1975: 111-118). The fault runs parallel to the grain of the Siwalik Hills, in the northwest-southeast direction.

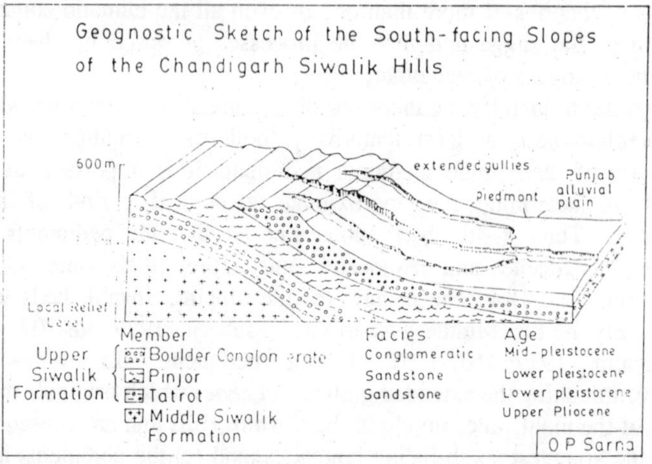

FIG. 19.6: Geognostic Sketch of the South-facing Slopes of the Chandigarh Siwalik Hills.

Although all the theories of pediment formation can successfully explain the processes operating in different climatic contexts and combinations are still more effective some theories, at least, can be discarded from their application to particular contexts. The exhumation theory is simply not applicable to the Punjab piedmont pediments as there are no evidences for deep surface and subsurface rotting and their subsequent exhumation by removal. Since the Boulder Conglomerates provides the first surface on which the pediment forming processes could operate and the pediments are presently sitting on them, it is clear that the pediments have not been created through the removal of an younger bed and the weathered material overlying them. The Boulder Conglomerate is the youngest bed of the Siwalik formation. Thus, the process of exhumation did not operate here. It may also be pointed out that during a fairly long period during the later phases of the Upper Pleistocene and well within the Holocene arid climatic conditions continued entirely eliminating the possibility of deep surface and subsurface rotting.

The exhumation theory applied by a large number of geologists and geomorphologists (Oberlander, 1974: 849-875; Ruxton, 1958: 353-377; Ollier, 1960: 137-148; Tuan, 1959: 1-140 and Budel, 1957: 201-228) to the semi-arid and arid as well as Savannah regions was specially directed on the pediments which are genetically related and locationally attached to the inselbergs and tors. Since the inselberglandscape or tor landscape itself was considered as the result of deep surface and subsurface rotting and their subsequent removal by sheetflow, the process of exhumation, it was suggested that the pediments were also the exhumed surface. Now, in the Punjab piedmont zone the pediments are attached to the Siwalik Hills and their hogback ridges and their south-facing slopes and nowhere to inselbergs and tors. Thus again the exhumation theory cannot be applied to the Punjab piedmont pediments.

Considering the context of their present occurrence one could suggest that the pediments need a composite theory for their comprehensive understanding. But this composite theory has to be a bit more comprehensive than the composite theory proposed earlier (Bryan, 1935: 765-775). The Bryan composite theory involves lateral planation by stream, backweathering, sheet flow, removal of detritus by sheet flow, sheet wash, diffuse flow, laminar flow, and turbulent unconcentrated flow and was formulated by combining the theories of lateral planation, backweathering and recession, and sheet flow. Essentially two kinds of flow, sheet and turbulent, and two kinds of basic processes, lateral planation and backweathering are involved in this theory.

The lateral planation proposed in the composite theory is achieved only by stream but it is quite clear that this process, in order to create wide, perceptible, and long pediments, would have to operate either through wide swing of the stream laterally or through the lateral spreading of water in the form of strong over-bank flow of spillover water. This assumes that the streams are of moderate size and their discharge, when they are suddenly activated, is considerable. In such a situation the interfluves will completely disappear, an evidence that is completely missing in the Punjab piedmont region. Also it needs to be emphasised that the pediments thus created through lateral planation by rivers must be very extensive while the pediments of the study area are far smaller and there are no permanent, large streams in this tract.

Almost all the streams, *choes*, gullies, extended gullies and minor channels have permanent channels but ephemeral discharge. For most parts of the year the streams of all orders do not carry any water at all and when the discharge is full the steep gradient of the slope facet hardly allows any lateral swing and therefore lateral planation. On the other hand, by extending the arguments formulated by Wirthmann (Wirthmann, 1981: 165-204) against the climatic controls and emphasising the influence of rocks exposed on the surface on the pediment forming processes, it should be suggested that in the case of the Punjab piedmont pediments both the specific rock types and rainfall characteristics have been of crucial importance. The presence of the resistant Boulder Conglomerate on the surface and the innumerable gullies quite clearly indicate that both rock types and climate have been important during the Upper Pleistocene and Holocene times and at present in the formation of the pediments.

Eliminating the lateral planation by streams but not lateral planation as such the question of the origin of this process becomes quite evident. It is proposed here that this lateral planation is really a form of areal (sheet) erosion caused by overland sheet flow. Whether the sheet flow attains the dimensions of sheet flood in the down pediment sections or not is not really that crucial. It has been claimed that essentially areal (sheet) erosion is caused by bed-load progression in the downslope segment (Seuffert, 1981: 141-164). Seuffert explains that this kind of change leads to a polylinear convergence of the belt which is eroded by the running water. The main factor responsible for this erosion is the structure of the distribution of rain in space and time within the river catchment in question. Seuffert makes a very important suggestion that highly convectional rains with a very limited areal extent and a high and very varying intensity are most favourable for bed load progression downslope and therefore for areal (sheet) erosion (Seuffert, 1981: 141-164).

On the basis of the arguments stated in the preceding paragraphs we propose a different formulation on the formation of the Punjab piedmont pediments. This formulation or model is a modification of the Bryan composite theory to some extent but it does incorporate some original arguments as well.

To begin with we have to explain the recession of the hill-side slope and extension (widening and lengthening) and shaping of the pediment. At many, if not all, places the hill-side slopes are steep scarps. These scarps stand at the head of the pediments. Thus, essentially we are here dealing with the retreat of scarps. The shaping of the pediments mainly refers to the processes by which the lateral and longitudinal extents are produced. It is important to remember that a combination of processes effect the formation of the pediments.

There were clearly two stages through which the pediments were formed: tectonic followed by erosion. Thus both the hypogenic and epigenic processes were involved. In the first stage, a very brief one, a linear fault of modestly regional scale ran through the tract of the Boulder Conglomerate that stretches along the southern edges of the south flanking slopes of the Siwalik Hills (Fig. 19.3). The faulting that took place here caused a downward displacement of a narrow strip of the Boulder Conglomerate toward the piedmont and upward movement of the remaining part toward Siwalik Hill southern flanks. In the initial stage, therefore, the junction between the lower and upper strip blocks of the Boulder Conglomerate was initiated by faulting. It was a part of the neotectonic movements which have been widely documented (Nakata, 1975: 111-118). This junction lead to the creation of the geological context in which the pediments later developed, the junction angle of the pediment with the scarp, and the topographic slope of the scarp.

The scarp retreat was essentially achieved by planar erosion. A continuous deep notch was horizontally cut into the junction by several gullies jumping from the hill-side slope (corresponding to the upthrown strip block of the Boulder Conglomerate) on to the surface that would later become the pediment (corresponding to the downthrown strip block of the Boulder Conglomerate). The long notches are thus located at the base of the scarp. However, these notches are very different from those located at a higher site and proposed in the model of scarp retreat through the enlargement of the notches inside the base, undermining of the foundational support of the scarp, and steep rock-fall leading to its retreat (Twidale and Milnes, 1983: 325-343). In the pediments of the Study Area the scarp retreat was essentially achieved by gullies, some independent and some joining to form thin sheets (Plates 19.2, 19.3, 19.9 and 19.11). It was a combination of gully erosion and sheet erosion that helped in the recession of scarp and the creation of a particular junction angle and a topographical slope that at most sites is concordant with the structural dip. The gullies are concentrated mostly on the Boulder Conglomerates themselves and disappear on their entry into the piedmont alluvium, at the point were some of them integrate to form larger *choes* (Fig. 19.6).

During the early periods of the Holocene arid and semi-arid conditions prevailed in the Punjab piedmont zone. Several scholars from different disciplines have adduced massive evidence for the period of dessication during the Holocene (Wissman, 1956: 278-303; Zeuner, 1953: 242-253; Bhatia and Khosla, 1967: 507-509 and Mohapatra, 1979: 107-135). The dessication was mainly characterised by a combination of thundershower and convective downpour and far less the advective-orographic variety. The thunderstorm rainfall, essentially sharp and spotty shower, is highly characteristic of semi-arid (BS), arid (BW), and modified monsoonal (Cwg) climatic regimes. The intensity of rain is very high indeed leading to the generation of rapid sheet flow and sheet wash on slopes built of rather resistant, impervious, and unresponsive rock surfaces having a sparse vegetation cover. The Boulder Conglomerates of the south-flanking slopes of the Siwalik Hills answer this description of ideal conditions.

The second stage in the pediment formation begins with the generation of climatically generated processes (Fig. 19.7). The sharp spotty shower coming down from thunderstorms and strong convective ascent associated with cumulonimbus clouds of the summer monsoons generates diffuse run-off and sheetwash. The innumerable rills and gullies create a mesh and sheet of water on the pediments (Plate 19.12). Together, the rills and gullies and the sheet erosion effect the parallel retreat of the scarp developed on the beds of the Boulder Conglomerate. The junction is cleared off the grus by the sheetflow and gullies which bring them down after eroding the scarp slope. The sharpness of the junction is striking. There is no doubt that colluvium and grus through *in situ* weathering are produced during the long period of dry months of the year but the accumulation is prevented by their removal by sheetwash and gullies during the monsoon months (Plate 19.3). As the sheetwash moves down its velocity and turbidity increase and it excavates into the bedrock a ramplike erosion feature, the pediment which extends downslope corresponding with the outcropping of the bed rocks. It needs to be noted that in the Punjab pediments there are no talus or grus cones at the foot of the scarp.

Conclusions

The Punjab piedmont pediments are formal homologues of their Rajasthan counterparts but not their climatic analogues. The Punjab pediments are not the integral elements of the classical King model as they are backed by a constant slope with free face and convex slopes missing. Their origin is located in the later phase of the Upper Pleistocene and the early phase of the Holocene but is not associated with lateral planation by streams. Two processes occurring in temporal sequence lead to their genesis: faulting and sheet erosion. A recent regional fault running through the Boulder Conglomerate strip lead to the relative, displacement of two linear blocks, the higher provided the setting for the evolution of the scarp and the lower was the ground on which developed the pediment. Sheetwash and backweathering generated by high intensity, spotty, thundershowers were then dominant pediment forming processes. Although the climaxial stage of evolution was attained and passed in the late Holocene there has been a slow, rather imperceptible change in the pediment in the later geological periods. Presently, the pediments appear, more or less, dead features, fossil in character. The pediments are features of both climatic geomorphology and structural geomorphology.

Acknowledgements

The author is deeply grateful to Dr. Baldev Singh Karir (Department of Anthropology, Panjab University) for help in conducting the field investigations and to Mr. O.P. Sarna for drafting the sketches and maps.

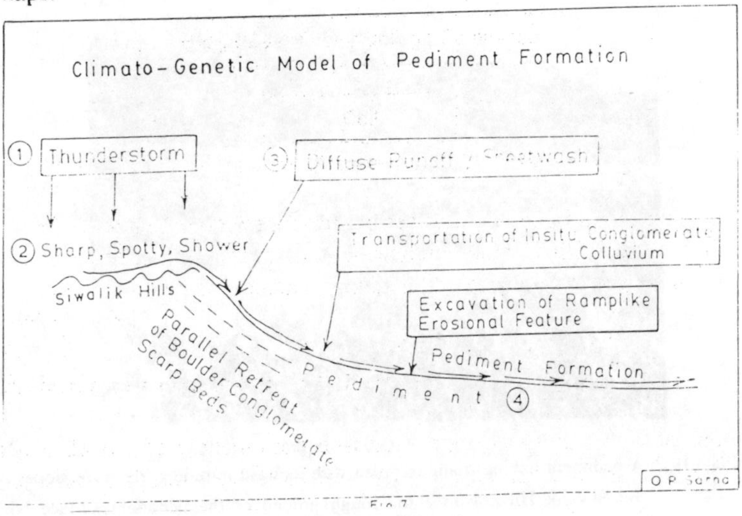

FIG. 19.7: Climato-genetic Model of Pediment Formation.

Plate 19.1: A panormamic view of the Siwalik Hills and the pediments stretching out toward the middle ground. Note the hogback ridges.

Plate 19.2: A pediment having triangular plan with its head intruding the scarp slopes of the Siwalik Hills. Notice the rough ground of the pediment surface with scattered tufts of grass.

Plate 19.3: Note the sharp junction between the dipping beds of the Siwalik Hills and the pediment below them. Recession of the scarp is indicated by the remanants of the Boulder Conglomerate beds.

Plate 19.4: An extensive stretch of pediment clearly revealing the proximal segment in the middle ground and the distal segment in the foreground. Note the bold scarp behind the pediment head.

Plate 19.5: A scarp remanent at the head of the pediment. The fairly steep pediment surface has been cut by a low, artificial wall.

Plate 19.6: A thin cover of grus, very recently chemically and mechanically weathered. Note the cover of *bhabar* (*Elephanta* Sp.) grass. The gravels and pebbles have been derived from the Siwalik Beds.

Plate 19.7: An exposure of Boulder Conglomerate beds at the terminal of a pediment. The layers of clay with earth-pillars boldly etched are separated by pebble-gravel thin interstratified layers.

Plate 19.8: Terminal of the pediment toward the piedmont alluvium. Notice the tufts of grass on a surface covered with pebbles and gravels.

Plate 19.9: A closer view of the hogback slope and the pediment stretching to the left.

Plate 19.10: A section through the thick Boulder Conglomerate beds in the middle segment of the pediment. Note the interstratified layers of pebbles and gravels.

Plate 19.11: Note the longitudinal gully extending over the pediment in the middle ground and separating one pediment from the other. The linear depression marks the path of the gully.

Plate 19.12: The pediments are enclosed by solid black lines. Note the sites of the pediments and innumerable extending gullies carving the southfacing slopes.

REFERENCES

Beaty, Charles B. (1974): 'Debris Flows, alluvial fans, and a revitalised catastrophism', *Zeitschrift fur Geomorphologie*, Vol. 21, December, pp. 39-51.

Bhatia, S.B. and Khosla, S.C. (1967): 'Sub-recent Ostracodes from Southern Punjab, Tahsil Bhiwani, District Hissar', *Research Bulletin of the Panjab University*, Vol. 18, Nos. 3 and 4, pp. 507-509.

Blackwelder, E. (1931): 'Rock cut Surfaces in Desert Ranges', *Journal of Geology*, Vol. 20, pp. 442-450.

Bradley, W.M. (1940): 'Pediments and Pedestals in Miniature', *Journal of Geomorphology*, Vol. 3, pp. 244-255.

Bryan, K. (1932): 'Pediments Developed in Basins with through Drainage', *Bulletin, Geological Society of America*, Vol. 43.

——— (1935): 'The Formation of Pediments', *Report of the 16th International Geological Congress*, Part 2, pp. 765-775.

——— (1935-36): 'Processes of Formation of Pediments at Granite Gap, New Mexico', *Zeitschrift fur Geomorphologie*, Vol. 9, Part 4, pp. 125-135.

——— (1940): 'The Retreat of Slopes', *Annals, Association of American Geographers*, Vol. 30, pp. 254-268.

Budel, J. (1957): 'Die Doppelten Einebnungsflachen in den Teuchten Tropen', *Zeitschrift fur Geomorphologie*, Vol. 1, pp. 201-228.

Chatterji, P.C., Singh, Surendra, and Qureshi, Z.H. (1978): 'Hydrogeomorphology of the Central Luni Basin, Western Rajasthan (India); *Geoforum*, Vol. 9, pp. 211-224.

Chaudhri, R.S. and Gupta, V.J. (1969): 'A Contribution to the Geology of Haritalyangar and Adjoining Region', *Bulletin, Indian Geologists' Association*, Vol. 2, Nos. 3 & 4, pp. 102-104.

Chaudhri, R.S. (1974): 'Structural features of the Lower Tertiary Rocks of Simla Hills, Northwestern Himalaya', *Journal of Indian Geosciences Association*, Vol. 17, pp. 47-51.

——— (1976): 'The Problematic Stratigraphical Aspects of the Cenozoic Sediments of North-Western Himalaya — A Critique', *Himalayan Geology*, Vol. 6, pp. 221-239.

Cooke, R.U. and Mason, P.F. (1973): 'Desert Knolls Pediment and Associated Landforms in the Mojave Desert, California' *Ravue de Geomorphologie Dynamique*, Vol. 22, No. 2, pp. 49-61.

Davis, W.M. (1938): 'Sheet floods and Stream floods', *Bulletin Geological Society of America*, Vol. 49, pp. 1337-1416.

Denny, C.S. (1967): 'Fans and Pediments', *American Journal of Science*, Vol. 265, pp. 81-105.

Dury, G.H. (1972): 'A Partial Definition of the term pediment with field tests in humid climate areas of Southern England', *Transactions, Institute of British Geographers*, Vol. 57, pp. 139-152.

Fair, T.J. (1948): 'Hillslopes and Pediments of the semi-arid Karoo', *South African Geological Society Journal*, Vol. 30, pp. 71-79.

Gansser, A. (1964): *Geology of the Himalayas*, London: Interscience Publishers.

Gaur, Rajan and Chopra, S.R.K. (1984): 'Taphonomy, Fauna, Environment, and Ecology of the Upper Siwaliks (Plio-Pleistocene) near Chandigarh, India', *Nature*, Vol. 308, No. 5957, pp. 353-355.

Gill, G.T.S. (1985): 'Petrology of the Siwalik Sequence Exposed Between the Rivers Ghaggar and Markanda, Northwestern Himalaya', *Publications, Centre of Advanced*

Studies in Geology, Panjab University, New Series, Vol. 1, pp. 43-58.

Kar, Amal, Singh, S., and Ghose, Bimal (1977): 'Geomorphological Characteristics and Palaeoclimatic Significance of Rhyolite Pediments in Jodhpur District, Western Rajasthan', *Man and Environment*, Vol. 1, pp. 16-20.

Kar, Amal, Singh, S. and Kaith, D.S. (1979): 'Relations of Debris Size and Slope with Distance Along two Rhylite Pediments in the Arid Environment of Rajasthan', *Annals of Arid Zone*, Vol. 18, No. 4, pp. 236-245.

Kar, Amal (1984): 'Morphology and Evolution of Some Sandstone Pediments in the Rajasthan Desert, India', *Geographical Review of India*, Vol. 46, No. 1, pp. 67-74.

——— (1984): 'Lithological Control on Pediment Form and Morphology : Some Investigations in the Rajasthan Desert, India', *Geographical Mosaic*, (ed.). S.C. Mukhopadhyay, pp. 21-27.

Kesel, R.H. (1973): 'Inselberg Landform Elements : Definition and Synthesis', *Revue de Geomorphologie Dynamique*, Vol. 22, No. 3, pp. 97-109.

King, L.C. (1957): 'The Uniformitarian Nature of Hillslopes', *Transactions, Edinburgh Geological Society*, Vol. 17, pp. 81-102.

King, Lester (1962): *The Morphology of the Earth*, Edinburgh : Oliver & Boyd; pp. 698-699.

Lumley, H.D., Miskovsky, J.C.; and Miskovsky, J.R. (1973): 'Depots Du Riss Et Du Riss-Wurm Dans Le Midi Mediterrannean', *Le Quaternaire*, pp. 62-67.

——— (1973): 'Le Wurmien Ancient Dans Le Midi Mediterrannean Francais D'Apres L'Etude Des Depots De Grottes Et Abris Sous Roche', *Le Quaternaire*, pp. 79-89.

Mc Gee, W.J. (1987): 'Sheeflow Erosion', *Bulletin, Geological Society of America*, Vol. 8, pp. 87-122.

Mohapatra, G.C. (1979): 'Pattern of Cultural Growth in Prehistoric North-West India', in *Early Man in North-West India*, (ed.) S.R.K. Chopra, pp. 107-135.

Mukerji, A.B. (1976): 'Choe Terraces of the Chandigarh Siwalik Hills, India : A Morphogenetic Analysis', *Revue de Geomorphologie Dynamique*, Vol. 26, No. 1, pp. 1-19.

——— (1979): 'Early Man's Habitat in the Siwalik Hills', in *Early Man in North-West India*, (ed.) S.R.K. Chopra, pp. 98-107, New Delhi : Allied Publishers.

Nakata, Takashi. (1975): 'On Quaternary Tectonics around the Himalayas', *The Science Report of the Tohoku University*, Vol. 25, No. 1, pp. 111-118.

Oberlander, T.M. (1974): 'Landscape Inheritance and the Pediment Problem in Mohave Desert of Southern California', *American Journal of Science*, Vol. 274, pp. 849-875.

Ollier, C.D. (1960): 'The Inselbergs of Uganda', *Zeitschrift fur Geomorphologie*, Vol. 4, pp. 137-148.

Ruxton, B.P. (1958): 'Weathering and Subsurface Erosion in granite at the Piedmont Angle, Balos, Sudan', *Geological Magazine*, Vol. 95, pp. 353-377.

Sahni, M.R. and Khan, E. (1969): 'Stratigraphy, Structure, and Correlation of the Upper Siwalik East of Chandigarh', *Journal, Palaeontological Society of India*, Vol. 4, pp. 61-74.

Seuffert, O. (1981): 'Zur Theorie der Fliesswasserosion', *Geookodynamik*, Vol. 2, pp. 141-164.

Singh, Surendra and Kaith, S. (1971): 'Geomorphological Analysis of Landform Features of Pali Development Block From Aerial Photographs', *Indian Geographical Journal*, Vol. 10, pp. 50-59.

Singh, Surendra. (1972): 'Geomorphology and Climatic Changes During Quaternary Period in Rajasthan Desert', *Recent Researches in Geology*, Vol. 5, pp. 121-130.

———. (1977): *Geomorphological Investigation of the Rajasthan Desert*, Jodhpur, Central Arid Zone Research Institute, pp. 6-7.

Tandon, S.K. (1974): 'Litho-control of some geomorphic properties : An Illustration from the Kumaun Himalaya, India', Vol. 18, No. 4, December, pp. 460-471.

Twidale, C.R. (1967). 'Origin of the Peidmont Angle as Evidenced in South Australia', *Journal of Geology*, Vol. 75, pp. 393-411.

Twidale, C.R. Bourne, J.A., and Smith, D.M. (1974): 'Reinforcement and Stabilisation Mechanisms in Landform Development', *Revue de Geomorphologie Dynamique*, Vol. 23, No. 3, pp. 115-126.

Twidale, C.R. and Milnes, A.R. (1983): 'Slope Processes Active Late in arid Scarp Retreat', *Zeitschrift fur Geomorphologie*, Vol. 27, No. 3, pp. 257-264.

Tuan, Y.F. (1959): 'Pediments in Southeastern Arizona', *University of California, Publications in Geography*, No. 13, pp. 1-140.

Vishnu-Mittre, (1965): 'Floristical and Ecological Reconsiderations of the Pleistocene Plant Impressions from Kashmir', *The Palaeobotanist*, Vol. 13, No. 1, March, pp. 308-328.

Wirthmann, A. (1981): 'Taler, Hange Und Flachen in Den Tropen', *Geookodynamik*, Vol. 2, pp. 165-204.

Wissman, H.V. (1956): 'On the Role of Nature and Man in Changing the Face of the Dry Belt of Asia', in *Man's Role in Changing the Face of the Earth*, (ed. W.L. Thomas), Chicago: Chicago University Press, pp. 278-303.

Zeuner, F.E. (1953): 'Das Problem der Pluvialzeiten', *Geologische Rundschau*, Vol. 41, pp. 242-253.

20

Study of Some Aspects of Applied Geomorphological Features of Ladakh and Trans-Himalayan Region

R.C. PATHAK

Introduction

THE VERITABLE, intricate mosaic of Himalayas have a vast reservoir of glaciers and water resources. One of the second largest southern glacier (Siachen) in Nubra-Shyok Valley also exists around Karakoram Ranges. Most of the Himalayan glaciers are retreating and so is the case of the one studied in the present work. Some of the curious and unique geomorphological features of interest to the geomorphologists, have been observed during the field reconnaissances and suitably presented.

The geological regionalisation of the Ladakh and Trans-Himalayan regions, explain the Indus and Shyok-Nubra Valleys as occupying the discontinuous steeply thrusted suture (or paleaeosubduction) zones with ophiolitic melanges. The geomorphological typology and evaluation of the terrain describes the region as an ectropic zone with retarded valley cutting. The region presents the zones with perched, barren, moonland such as land pattern which is devoid of vegetation due to cloud and air-mass barrier ranges of Greater Himalayas. Xerophytic shrubby bushes and stunted vegetal cover are observed along the valleys of Indus and Shyok-Nubra. Ladakh complex is a high altitude winter cold desert, a morphoclimatic ectropic zone where surface transformation by geomorphological processes are extremely retarded. The study area geomorphologically can be suitably delineated into four landform pattern:

(a) Shyok-Nubra Valley
(b) Ladakh Massif
(c) Indus Valley
(d) Zaskar Range

During the field study of the glacier, glacial and periglacialareas, emphasis has been laid to analyse the stratigraphic layers (dirt bands) of the glacier and their deformation pattern; recession or retreat of glacier snout and glacial-thermo-fluvial erosion phenomenon. Folds, cracks, crevasses and other features developed in glaciers by flow characteristics closely resemble structures observed in metamorphic rocks as ice and rock are both polycrystalline solids and more or less deform in similar ways. It could be observed that to quite a degree of resembleness rock mechanics can be applied to glacier behaviour under some uniform constant conditions. Thus, studying the origin of structures in glaciers can shed some light on fundamental geological processes. Apart from the above some of the land sculptured into an intricate pattern of gullies, small ravines which give rise to interesting structures like earth pillars development have also been studied.

The above interesting findings have been analytically presented in the succeeding paragraphs and attempt have been made to logically deduce their causative factors.

Field Study

Extensive field visits have been carried out and observations made. Five years data (1984-85 to 1988-89) and ground truths have been verified and conformed. The general area of study around Leh-Kargil road and Nubra-Shyok Valley has been depicted in Fig. 20.1. In Western Himalaya, the longest and largest glacier in Nubra-Shyok Valley (Siachen) has also been shown in Fig 20.1. There is a group of Southern glaciers on Southern side of Karakoram, which provide potential scientific database for research work and has been analysed. One such large southern glacier has been selected for study (Latitude 35-38° and Longitude 75-78°). The typical glacier end-moraine (front facing) along with existing topography has been photographed in the Fig 20.2. It has been observed that the particular glacier is debris-covered with a layer of about 30 cm morainic dust towards terminal moraine and snout areas. The debris thickness gradually recedes towards accumulation zone of the glacier.

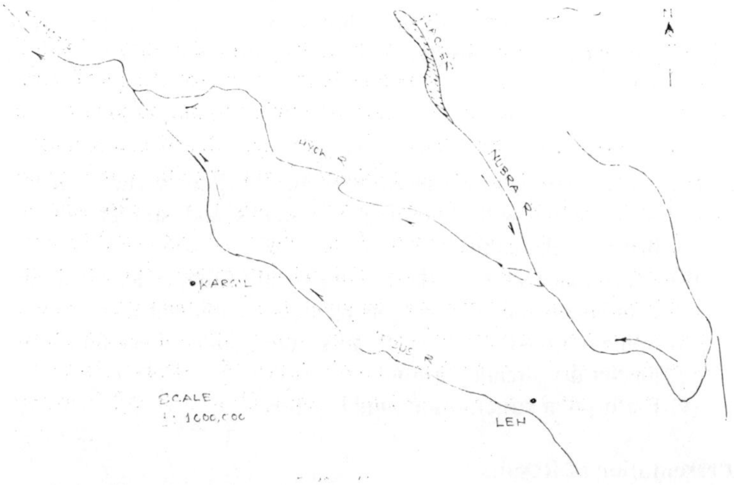

FIG. 20.1: Nubra-Shyok Valley.

The maximum temperature (average) in Nubra Valley has been recorded as 17.5°C during August month and minimum -7.7°C (average). The lowest temperature at this altitude, 3550m, is found to be -20°C. The maximum melt-rate of glacier is also found to be during August month. Around Leh area the maximum temperature at altitude 3400m is approximately 25°C and minimum –.5°C. The maximum wind speed in the Nubra Valley has been recorded as 42kph and practically no rains or snow precipitation in these regions of Ladakh is a common feature. In Leh area only 37 mm of rain has been recorded per annum (Pathak, 1988).

Most of the Himalayan ranges belong to the relatively recent Alpine orogeny which lie closer to equator and tectonically younger. The Himalayas are full of folds and faults and many thrusting action in continuum. The geological regionalisation of the Ladakh region explains the Indus and Shyok-Nubra valleys as occupying the discontinuous, steeply thrusted suture (or palaeosubduction) zones with ophiolitic melanges. Also, the geomorphological typology of the area explains the Ladakh region as an ectropic zone with retarded valley cutting. The vegetal cover, topography and the environmental aspects of the four landforms studied viz. Shyok-Nubra Valley, Ladakh

Massif, Indus Valley and Zaskar Range are suitably presented in the text.

Some of the following geomorphological features obtaining in the study area have been observed, of which the analytical details are described in this paper:

(i) Retreating phenomenon of glacier.
(ii) Close-up of glacier snout showing the high vertical blue ice walls along with tumbling ice blocks and stratigraphic dirt bands or glacial till.
(iii) A peculiar feature of thermo-erosional cutting of ice creating an overhang niche all along glacier side longitudinally.
(iv) Some glacio-fluvial round pits configurations on the bed of glacier dry stream portion.
(v) Earth-pillar structures around Leh area.

Presentation of Results

The analysis carried out and geomorphological study of glacial and periglacial areas of Himalayas have been presented in the succeeding paragraphs.

Geomorphological Regionalisation of Ladakh Complex

It is observed that the regional individuality and consistent features of the terrain of Ladakh region are the convoluted mass of mountains and linear valleys on the one hand and the extreme aridity on the other. The region presents Ectropic Zone with perched, barren moon land like land-pattern which is devoid of vegetation due to cloud and air mass barrier ranges of Greater Himalayas. Xerophytic shrubby bushes and stunted scarced vegetal cover are most frequent along the valleys of Indus the Shyok-Nubra and other riverine terrain.

The environmental characteristics taken together, topographically, Ladakh is a high altitude winter-cold desert i.e., a morpho-climatic ectropic zone where surface transformation by geomorphological processes are extremely retarded (Pal, 1988; Washburn, 1979; Wadia, 1961 and Pathak, 1988). The study area is the loftiest inhabited region in the world as perhaps having the most singular morpho-climatic in the world, where joint effect of elevation, extreme aridity of the climate marked by winter cold characteristics and diurnal range of temperature, perch everything and produce a perched landform – a desert of bare crags of granite dust and lofty mountains of different hue and shade (Rizvi, 1983). The Ladakh complex has been delineated into the following four landform units, which are described subsequently in brief:

(a) Shyok-Nubra Valley
(b) Ladakh - Massif
(c) Indus Valley
(d) Zaskar Range

Shyok-Nubra Valley

The area is constituted by Shyok and its tributary, Nubra (Colbeck, 1980). The valley lies in north of 34° 30′ N and covers an area of about 5076 sq km. The area lies in the midst of mountain ranges of Karakoram, Ladakh and Saltoro. The molasse and flysch units of the Shyok-Nubra Valleys are mainly exposed in the Saltoro Hills bounded by the Nubra and Shyok rivers. A thick heterogeneous sequence of basalt and andesite rocks forming the type section of the Shyok volcanics, constitute the floor for the Saltoro Molasses, Flysch and ophiolitic Melanges. The Shyok-Nubra Valley (3000m) is a classic example of U-shape glaciated valley under the high ridges of Karakoram and Saltoro (6620m) with a net relief of 3620m. Various types of sediment/deposit exist such as, fluviatile, glacial, fluvio-glacial, sluch-flowed or solifluction debris. The predominantly coalescence of fans form valleys sides, are the products of an environmental transitional from glacial to fluvial. These are termed by Eyles (1983) as paraglacial phenomenon. In the Alpine-Nival zone of Shyok-Nubra valleys, complexly fissured and jointed rock-head exhibit rock slopes in an over-steepened state of glacier retreat.

Ladakh-Massif

Ladakh-Massif more commonly known as Ladakh-Grani-toid upland is characterised by an intrusive grani-toid body of Ladakh Range which extends all along the entire length of the Indus suture zone in Ladakh as a NW-SE trending batholytic massif. Height rises from 3000m in Suru Valley on the west and form 2000 to 3000m in Indus Valley on the east to more than 6000m at the top of the Ladakh Range giving a relief amplitude of 3000 to 3200m. It is observed herein that the geomorphology of the Lakakh Range is petrographically influenced by the foliated and intensely jointed bedrock of granite which control the drainage lines on both flanks of Ladakh Massif.

Indus Valley

This valley appearing as a wide furrow, running parallel to the west-NW to east-SE alignment of the Himalaya is located between the Ladakh gran-toid massif and the sedimentary petrovariances of the Zaskar Range. Tectonically the Indus Valley occupies the Indus Suture Zone (like the Shyok-Nubra Valley occupying the Shyok-Suture Zone) in the South, the Zaskar thrust separates this palaeosubduction zone by south dipping thrust from the tectonic province of the Tethys Himalayan Zone and to the North, the south dipping Ladakh thrust separates from the Ladakh Magnetic Arc.

Zaskar Range

This region is a very good example of glacial shaped morphology and slush-flowed periglacial landforms. Zaskar differs from Ladakh as well as from Karakoram in the context of petrography and structural pattern. Suru, Zaskar and Indus are the main rivers of this area. In this region solifluction is a process as active as Cryoclastly and slush-flowed fans with boulders embedding are the most unique features, not found in other three regions of Ladakh (Fort, 1983). The morainic deposits and talus fans may be considered as juxta-glacial and pro-glacial cone-terraces built by snow-melt water. The Zaskar range belongs to the Cambrian system of folded sedimentaries including multicoloured lime-stones (of Fotu-La) and the exotic blocks of shelf carbonates which generate the steepest slopes of the area and contorted trellis drainage pattern is observed.

Glacier Retreat

The frontal view of the longest southern Nubra glacier has been shown in Fig. 20.2. The retreat of valley glacier marks have been shown by arrows in Fig 20.3. It is to be noted herein that most of the Himalayan glaciers are retreating in the present era. The observations during 1984-87 reveals a rate of recession of glacier to be around 1 to 2 metres per annum. The measurement has been taken on the snout by markers (Pathak, 1988).

FIG. 20.2: Frontal View of Study Glacier.

FIG. 20.3: Glacier Retreat Marks.

Stratigraphic Study of Glacier

Figure 20.4 represents various stratigraphic layers of glacier consisting of boulders till and morainic soil. The thickness of these dust or dirt-bands are observed to be 0.5m to 3m. The waviness of dirt bands proves the movement of glacier and deformation due to pressure mechanism of ice body (Hewitt, 1968; Paterson, 1983 and Pathak, 1988). The study of dirt bands can help in computing identification of rock-structural materials embedded and computing the age of glacier-mass.

FIG. 20.4: Dirt Bands of Glacier.

Thermal-Fluvial Erosion

In Figs. 20.5 (a) and 20.5(b), a typical thermofluvial erosion have been shown which is primarily due to thermo-karst and fluvial processes of glacial stream. The overhang niche portion of glacier ice (2 to 3m in depth) is annotated by the arrows. With the passage of time and due to fluvial action the cantilever portion keeps falling down into water due to its excessive weight and undercutting. A more or less similar feature has been reported developed at Ballast Brook, North West Bank Island Canada (French, 1979). Apart from the cantilever niche, herein, also it is observed that the glacier side slopes to 70° - 80°, which is almost vertical. This aspect suggests that geotechnically glacier-mass behaves like metamorphic rock type structures and some principles of rock-mechanics can be applied to the glacier study (Colbeck, 1980). However, the basic difference is that the rock deformation takes place in the earth's interior or in exposed condition and may require millions of years whereas ice deformation is much more rapid and can be observed morphometrically in the glaciers. Also, application of some of the methods and results of glacier flow analysis to problems of flow in the earth might be an improvement over the current treatments and practice followed in the earth science literature.

FIG. 20.5 (a)

Glacio-Fluvial Feature

A more peculiar glacio-fluvial features have been recorded in Fig. 20.6 where some are annotated by letters A, B and C. These are crater-like pits formed at the river sandy bed and are seasonal in nature and are

FIG 20.5 (b)

FIG 20.5: Fluvio-Thermal Erosion Features .

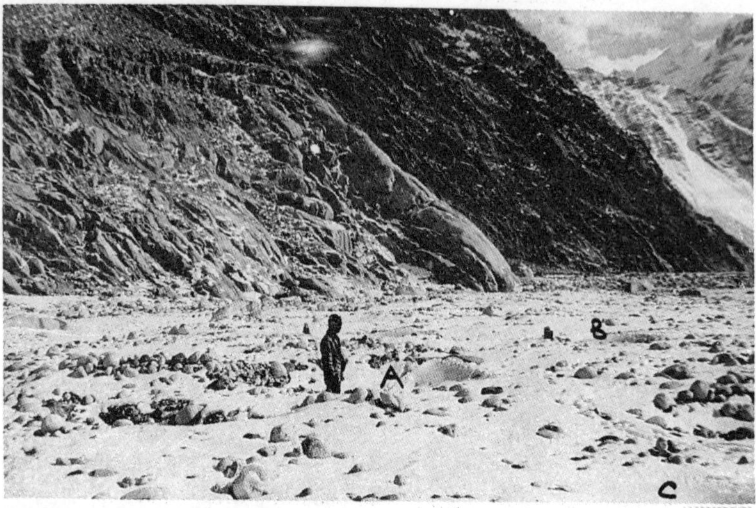

FIG. 20.6: Crater-like Features, A, B and C .

possibly due to ice chunks/blocks floating during high flood level period. The pits suggest the size of the ice body which ultimately has melted away. The ice mass and size of pits can help in computing dynamics of the running stream and mass-wastage of glacier body.

Earth-Pillar Structures

Apart from the above, some unique and curious structures such as earth-pillars sculptured into an intricate mosaic of gullies and small ravines separated by sharp spurs and buttresses, have been observed developed around Leh area (Altitude: 3400m) on the slopes of adjoining valley hills. The material constituting are mainly easily eroded conglomerates and boulder-clay and has possibly been found developed even in scanty rains and may be logically attributed due to strong winds and semi-arid cold weathering of this region.

Acknowledgements

The author is very much grateful to Director, R & DE (Engrs.), Dighi, Pune-15, for permitting to publish the paper. Author is also thankful to Scientists of SASE and DTRL, laboratories for essential help and discussion.

REFERENCES

Colbeck, S.C. (1980): "Dynamics of Snow and Ice Masses", Academic Press, (U.K.).

Eyles, N. (1983): "Glacial Geology - An Introduction for Engineers and Earth Scientists", Pergamon Press, USA.

French, H.M. (1976): "The Periglacial Environment", Longman, London (U.K.).

Fort Montique (1983): "Geomorphological Observations in the Ladakh area (Himalaya)", Quarterly Evolution and Present Dynamics, Hindustan Pub., Delhi.

Hewitt, K. (1968): "The Freeze-Thaw Environment of the Karakoram Himalayas", *Canadian Geog*. 12, pp. 85-98.

Paterson, W.S.B. (1983): "The Physics of Glaciers" 2nd Edition, Pergamon Press.

Pathak, R.C. (1988): "Study of Geotechnical and Applied Aspects of Himalayan Snow Bound Regions". Unpublished Ph.D. Thesis, Avadh University, (U.P.), India.

Pal, Saroj K. (1983): "Investigating the Relief Generations in Winter Cold Ladakh Region — A Study in High Altitude Geomorphologies", 72nd Ind.Sc. Congress, Lucknow (UP).

Rizvi, Janet (1983: "Ladakh-Cross-Roads of High Asia", Oxford University Press, Delhi.

Wadia, D.N. (1961): "Geology of India" - 3rd Edition, Macmillan, London, (U.K.).

Washburn, A.L., (1979): "Geocryology - A Survey of Periglacial Processes & Environments", Edward Arnold, London, U.K.

21

Some Geomorphic Observations in the Middle Reach of the Parvati River, Kulu District, Himachal Pradesh (India)

MANJARI BHATTACHARJI and V.A.V. RAMAN

PARVATI RIVER – a tributary of river Beas in the Kulu district of Himachal Pradesh rises from the Mantalai glacier near Pin Parvati pass and flows in a south-east to north-west direction till Manikaran from where it takes a south-west turn and flows down to join river Beas near Bhuntar.

The study is confined to a stretch of about 300 metres of Parvati river in its middle reach. Here the river passes through a deep gorge with an approximate depth of 25-30 metres near Gwachha. Just before it enters the gorge here it is joined by a large tributary – The Tos nala, on its right bank. After its exit from the gorge it is joined by the Gechu, Garthi, and Jigrai nala (Fig. 21.1). The left bank for a distance of about 150 metres upstream of the gorge stands as a precipitous cliff with a bare rock face.

Situated in the middle Himalayas, the study area falls within an altitudinal range of 2200-4200 metres above m.s.l. A series of interlocking spurs in this area breaks the ground into an intricate mountain mass with steep sided hills and narrow valleys. The topographic features are made by fluvial as well as glacial action.

Geologically the study area falls in the central crystalline nappe of the higher Himalayan zones comprising metasediments of different metamorphic grades. The metasediments include Dogra-slates, younger formations of uniform argillaceous sandy metasediments and their higher metamorphic equivalents, slates, sandstones, phyllites, mica-schists, biotite-quartzites and biotite gneiss. Lithologically seen

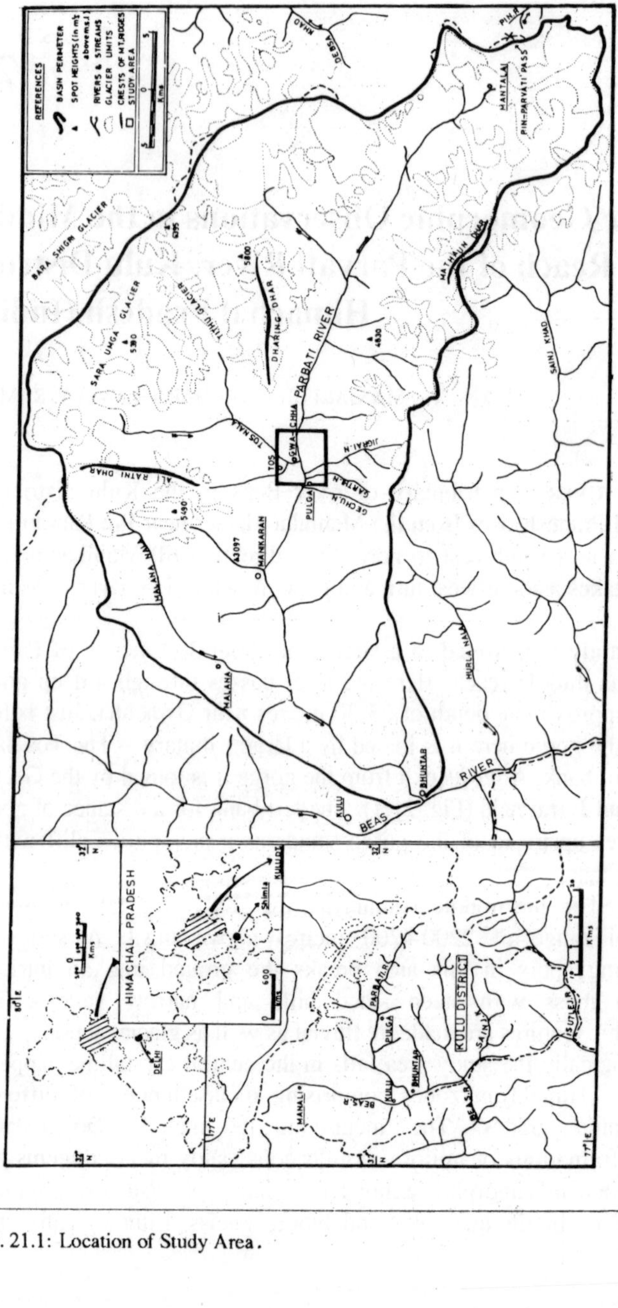

FIG. 21.1: Location of Study Area.

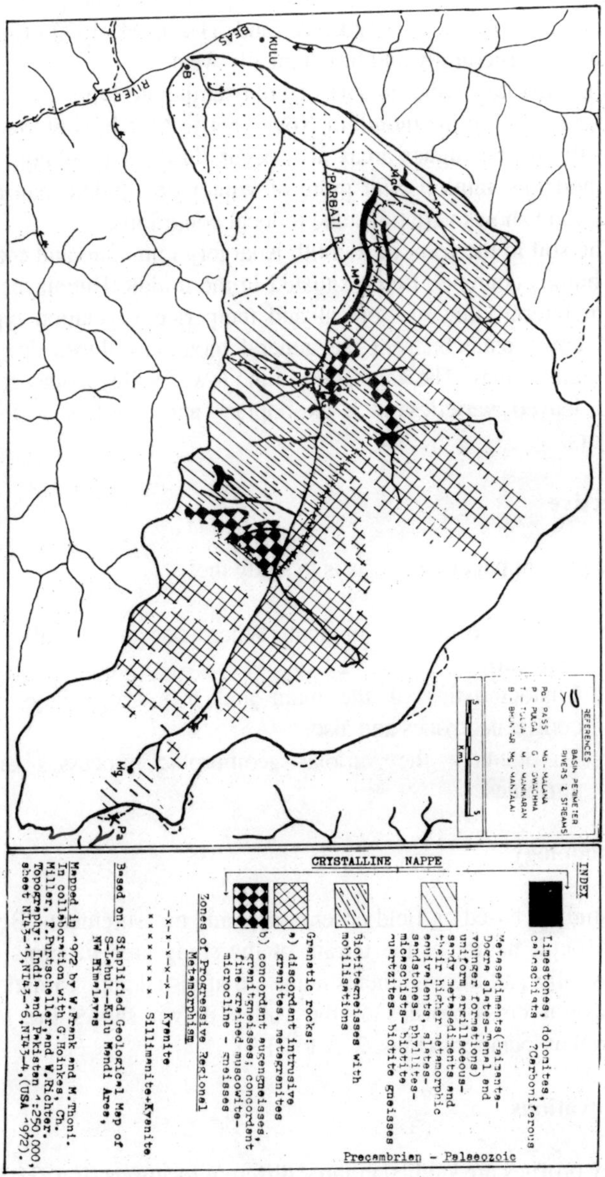

FIG. 21.2: Geology of the Area.

the area lies in a zone of progressive regional metamorphism-Kyantie, Kyanite+Sillimanite – (Frank. W. and Thoni. M, 1977) (Fig. 21.2). The area is structurally and lithologically weak.

The climate of the area is of cold temperate type. It experiences heavy snowfall in the month of November to March. However unusual snowfalls in the higher areas also occur during April-May. The area gets moderate rainfall with more of winter precipition than monsoon rainfall and shows variations due to local conditions.

The soil in the form of regolith is largely immature and coarse with high humus contents. Being situated in the middle Himalayas the area has an intermixture of tropical and temperate vegetation types. The forest cover extends from the general mountain slopes down to the inner valley sides. The vegetation in the lower valley slopes consists of broad leaved varieties such as poplars and conifers – Pines and Deodars.

Objective

In the present paper emphasis is to study the,

(a) Valley profile development (Where one of the valley sides is uniform),
(b) Characteristics of the main and tributary valleys near the confluence zones and also,
(c) To identify the various geomorphic processes and their attributes.

Methodology

The study is based on field observations and measurements. Because of the difficult nature of the terrain of the study area in terms of ready accessibility, all measurements made are of visual approximations.

Several cross-profiles of the valley drawn shows the form and material associations (Figs. 21.4 to 21.9).

Observations

Valley profiles are studied at three different locations (Fig. 21.3):

I. Upstream of the confluence of Parvati with Tos nala,

II. Along the Tos nala,
III. Along the Garthi nala.

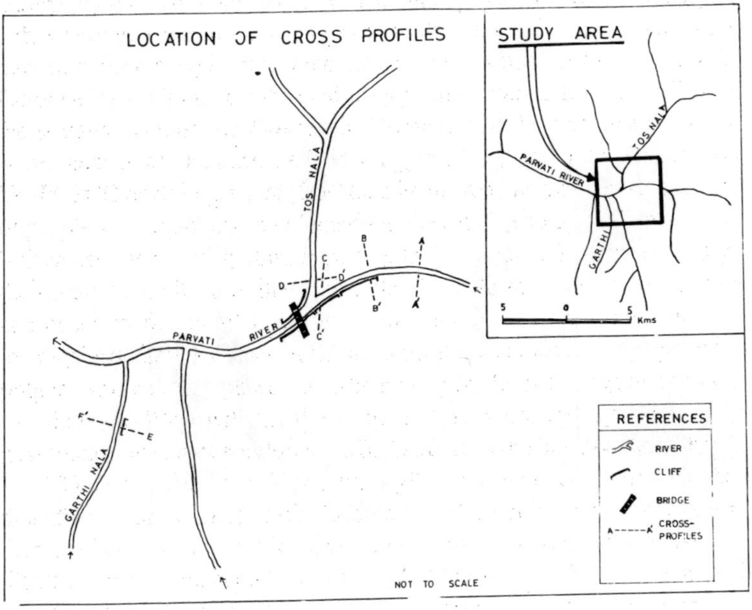

FIG. 21.3: Location of Cross Profiles.

Location I

Three profiles are studied and drawn from the centre of a bend of river Parvati up to the inflection point of the curve where the Parvati river meets Tos nala.

Profile A-A′ (Fig. 21.4)

This marks the centre of the bend of the river. Across the profile (A-A′) on the left side, the mountain slope descends steeply (35) down to the river bed without showing any clear break of slope. From this point the river bed spreads horizontally up to about 5-8 metres to the right bank which rises up with a conspicuous bluff of about one metre. The bluff marks the face of a gently sloping (10 to 15 towards river) surface of a width of 4-5 metres (Terrace) which continues on its upper side into the steep mountain slope.

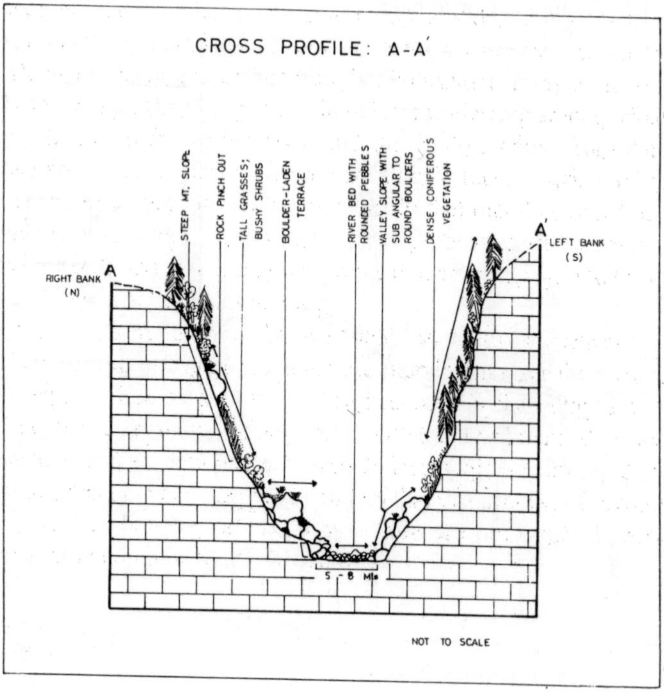

FIG. 21.4: Cross Profile A-A′.

The mountain slope on the left side is covered by dense vegetation with tall coniferous trees. The left bank portion is covered by huge sub-angular to rounded boulders. The river bed profile which is 5-8 metres in width is full of water rolled and rounded pebbles and cobbles with a boulder or two here and there. The gently sloping surface with a bluff on the right banks is in fact a surface of huge boulders with the gaps filled with gritty soil materials. The mountain slope on the right side unlike on the left side is covered by only tall grasses and bushy shrubs with a few rock pinch-outs.

Profile B-B′ (Fig. 21.5)

This is taken downstream at a distance of 120 metres from A-A′. The form of the profile differs from that of A-A′ in that it shows a precipitous cliff (25-30 metres) on the left side falling down to the river bed, which is about 8-10 metres in width, up to the right bank. The right bank and the associated mountain slope are of same configuration as of the right bank side (A) of profile A-A′. The cliff face on the left side down to its entire height is a bare rock face

Plate 21.1: View of the study area upstream of the confluence (right bank of the profile A-A').

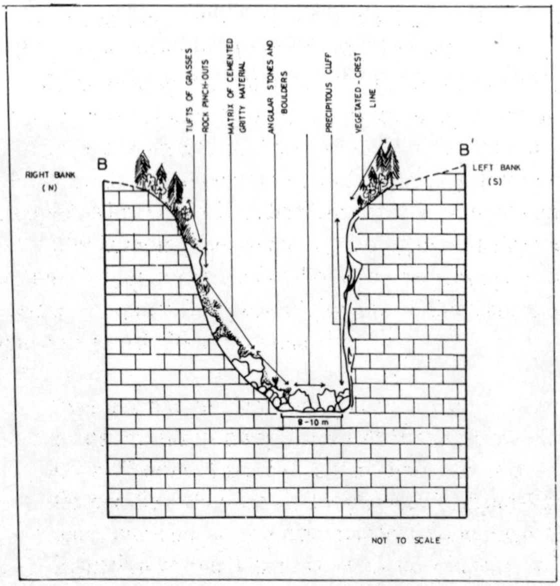

FIG. 21.5: Cross Profile B-B′.

The river bed is covered mostly by boulders and it is showing a profile in its bed, across a hung block of rock. The river bed grades up into mountain slope on the right side (B). It is covered by pebbles and cobbles sized angular stones with a few angular boulders interspersed here and there in a matrix of cemented gritty soil snow only tufts of grass here and there. There are trees high above this slope.

Profile C-C′ (Fig. 21.6)

It marks the upstream confluence point with Tos nala. The left side (C′) is a cliff face like in B-B′ with a notch at its base at the bed level. (There are a few notches along the same side downstream. Water is seen dripping down the face of the cliff in a continuous flow here and there and flow is marked by flutes and grooves). The bed here is

Plate 21.2: View of the gorge downstream of the confluence: note the notch at its base with an overhanging cliff on one side and a protruding bulge on the other side

wider than it is upstream along the other sections. This being the site for measuring the water discharge, the river bed appears to have been cleared. The right bank is covered by boulders of white and ochre yellow colour making a horizontal surface and it slopes up with a thick regolith. Higher level of this slope show broad leaved variety of trees. This profile marks the inflection point of the meander curve.

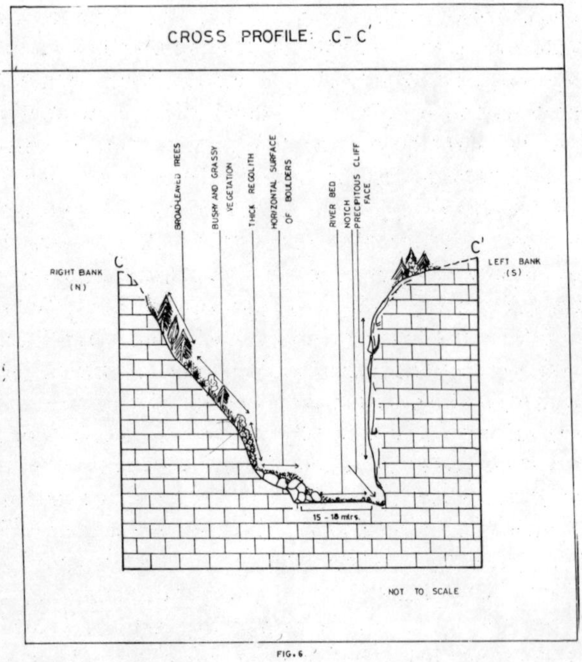

FIG. 21.6: Cross Profile: C-C'.

(i) A noticeable feature of the above three profiles is the abundance of materials (boulders, cobbles, pebbles, etc.) on the right bank and their conspicuous absence on the left bank.

(ii) The river flows through a gorge with a height of 25 metres approx. downstream from the confluence point with Tos nala, with a gentle curvature. The gorge shows a notch (all along) on the left side with the cliff over-hanging and on the right side the cliff shows a protruding bulge at its base (river bed). Here the left bank marks the undercut slope and the right bank marks the slip-off slope (nature of the bend is not reverse here, i.e., at the gorge site) (Fig. 21.7).

(iii) Three clear water marks are seen along the cliff face at heights of two metre, three metre, and four metre from the river bed. Bands of black, green and brown colour respectively marks these water line at the heights given above.

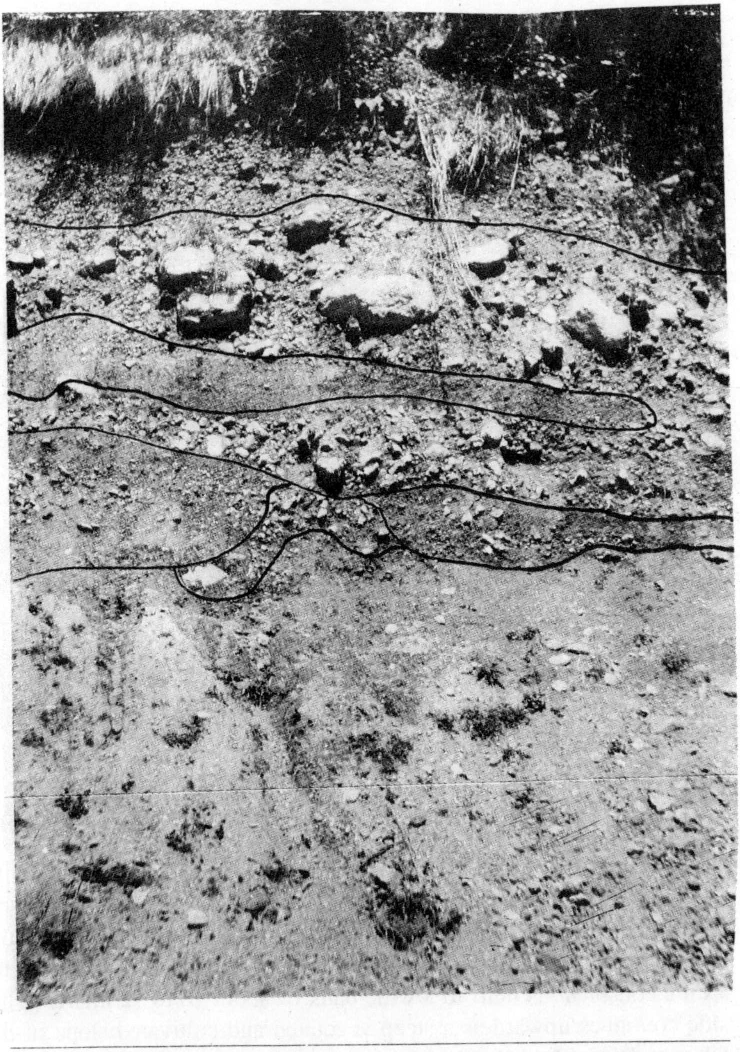

Plate 21.3: A View of the minor cliff on the right side mountain slope in the valley of the Garthi nala: Note the four horizons of different materials. These layers probably indicate periods of gravity associated creep and intervening periods of catastrophic deposition.

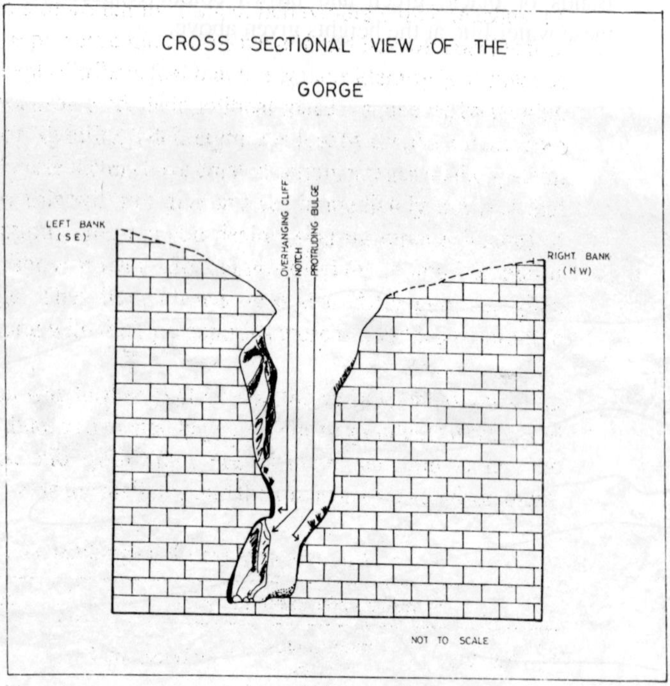

FIG. 7

FIG. 21.7: Cross Sectional View of the Gorge.

Location II

Cross Profile D-D' (Fig. 21.8)

This is taken across the Tos nala little upstream of its confluence with river Parvati. The right side falls as a steep vegetated valley side with an indistinct break of slope up to the river bed. From this point the river bed extends horizontally for a distance of 5-8 metres and rises up with a conspicuous bluff to a wide bank of about 10 to 12 metres. This side continues upward as a steep vegetated and cultivated slope of the Dharing Dhar (Range) thus giving the valley an asymmetrical form.

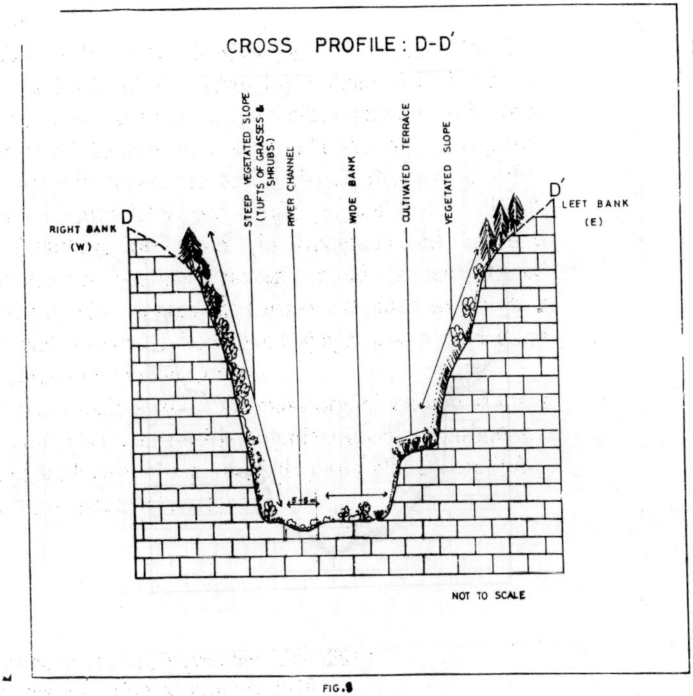

FIG. 21.8: Cross Profile: D-D′.

Location III

Cross Profile E-E′ Garthi Nala Profile (Fig. 21.9)

This profile shows steep valley sides with a good accordance in the breaks of slope on both sides. The left side comes down with a flight of terracettes up to the river bed which is covered by boulders. The river bed spreads horizontally to about 20-25 metres. The valley side on the right comes down steeply and turns into a cliff of about four metres in unconsolidated materials with four clear horizons of different materials. At the bottom, this minor cliff is a cone of scree whose toe comes up to the river bed.

The four horizons of meterials are:

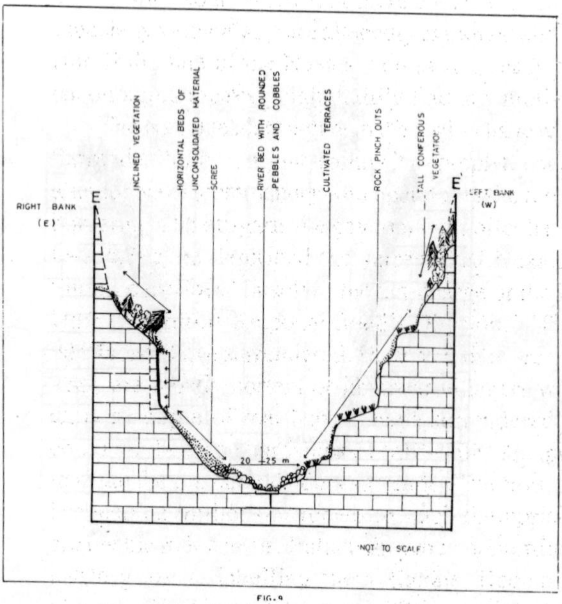

FIG. 21.9: Cross Profile E-E'

- Two layers of (one metre thickness) gritty material separated by a layer of angular pebbles and cobbles.
- The last layer of (one metre thickness) is fully gritty and gravely.

These layers probably indicate two periods of gravity associated creep and intervening periods of catastrophic deposition.

On the surface above this cliff the vegetation is slightly inclined from the vertical signifying thereby debris slide, and the whole material can at any time come down. On the other side of the river (left bank) the flight of terraces are all cultivated.

Inferences

A. In relation to the first objective the following is inferred,

(a) Pronounced asymmetry in the valley profile (Figs. 21.4-21.6). This may be attributed to;

(i) East-west alignment of the main river valley

The north-east south-west alignment of the valley makes the right side of the valley a south facing slope. As a result insolation and evapotranspiration is higher here. This in turn results in higher run-off and erosion than in the north-facing slopes, thereby rendering it more vulnerable to active mass-wasting. The weathering processes are also activated due to the associated micro-climatic influences.

(ii) Topographic plunge

With a glance of the contour pattern (Fig. 21.10), it can be seen that the northern side protrude to the southern side and they appear to be plunging to the south. On the other hand, left side mountain slope does not show any such projection to the north. As a result, the river is pushed to the south by the numerous spurs, while its southern movement is not counteracted by any topographic influence from the south. This gives an asymmetric form to the valley.

(b) Increase in the asymmetry in the downstream direction up to the confluence point, greatest being at the confluence. This is due to

(i) Entrance of tributary river (at the confluence site, profile C-C′).

The tributary (Tos-nala) supplies more debris to the Parvati river at the confluence site than the latter could remove. This leads to greater deposition of material on the right side of Parvati river resulting in a wider bank at the confluence. In the upstream direction the lesser asymmetry is a result of the decreased discharge and it's corresponding effect on the rate of mass-wasting on the right bank.

(ii) Spur truncation (nose of southwestern plunge) at the confluence by the Tos resulting in a wider bank here than upstream.

B. With respect to second objective it is inferred that there is an increase in the asymmetry of the tributary valley (Tos nala) in the downstream direction. This can be accounted for, by the

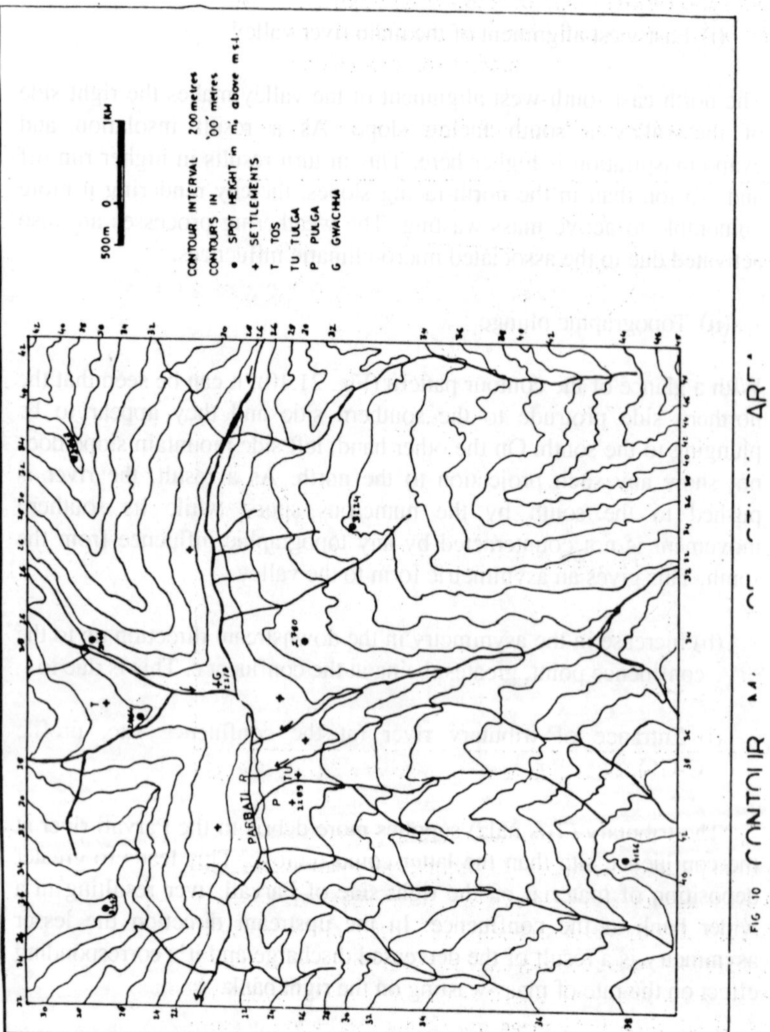

FIG. 21.10: Contour Map of the Study Area.

(a) Topographic projection:

The contour pattern of the Dharing -dhar, the left side mountain slope of the valley of Tos nala also show a topographic plunge to the east and south-east while the right bank shows a southeastward push to form the gorge. The former causes the river to embrace the right side mountain slope (Fig. 21.10).

(b) With reference to profile E-E′ (Garthi nala) it is inferred that the more or less accordant terraces or breaks of slope on either side of the valley is rather suggestive of a hitherto higher level of the river bed. Progressive down-cutting through fluvial action must have lowered the bed to the present level.

This valley is relatively more open in spite of the mountainous nature of the area. This can be attributed to the glacial action which must have carved out this ('U'-shaped) valley, presently under fluvial action (Fig. 21.9).

c. With respect to the third objective, i.e., to highlight some other geomorphic processes and their attributes, it is inferred that,

(a) With reference to location I. Point (ii) under profile (C-C′) described earlier,

The water marks show the height of water column in the valley (depths of flow in different times). The lowest mark at a height of two metres from the bad which is showing off as a two metre band of black colour obviously is the level which is frequently covered by water column. The next water line at a height of three metres with a band of one metre green colour (due to mosses) signifies periodic inundation. The fourth line with a difference of one metre band of brown colour seems to indicate rare peak flooding.

(b) The effect of the increased kinetic energy and/or increased discharge is seen in the form of a notch at the base of the gorge/cliff wall on the left side (refer profile (C-C′), at the confluence site. It would be pertinent to mention here that the distinctiveness or formation of such notches was perhaps aided and hastened by the curvature which characterise the gorge here.

(c) With reference to profile E-E of Garthi nala (Fig. 21.9) the inclined vegetation along the crest of the cliff on the right side of the valley is suggestive of the process of 'soil creep'.

Three horizons of materials of different nature along the small cliff (the right side mountain slope of the same valley – Fig. 21.9) offer evidence of changing intensity in the production and deposition of the same during different flood regimes. The layer of angular pebbles and cobbles may be suggestive of deposition in a more warm (humid/fluvial) spell and the layer of clay and silt during a cold spell indicating perhaps minor climatic fluctuations in the past.

Conclusion

The various geomorphic observations and the materials and their dispositions are indicative of neo-tectonics within this region.

Acknowledgements

The authors gratefully acknowledge Centre for Study of Regional Development, Jawaharlal Nehru University, New Delhi for sponsoring the field-study and having given the necessary facilities. They also convey their sincere thanks to

– Prof. (Dr.) K.S. Sivasami and Dr. P. Nangia of JNU for conducting them to the field,
– Dr. K.N. Prudhvi Raju and Prof. S.K. Pal of Jamia Milia Islamia and Delhi University respectively for having spared their valuable time during the preparation of the paper.

REFERENCE

Frank. W., Thoni. M., and Furtscheller. F., "Geology and Petrography of Kulu-South Lahaul area" in Himalaya - Sciences De La Terre. No. 28, ECNRB, France, 1977, pp.142-72.

22

Suitability of Image Processing Techniques for Identification of Geomorphological Features – A Case Study of Tirupati Area

P.S. DHINWA, JEBAKUMAR and K.L. MAJUMDAR

Introduction

REMOTELY-SENSED data contain a lot of information about various resources, e.g. agriculture, forestry, geology, geography, geomorphology, oceanography and landuse/landcover etc. But due to various reasons the data acquired by satellite do not occupy the full dynamic range and therefore the contrast is poor in the original image. From a user's point of view, the imagery with poor contrast is of very little use as it does not help the user in visual interpretation of specific feature, e.g. landform features. To help the user for better interpretability the image data should be subjected to various image enhancement techniques. Some of them are linear stretching, linear stretching with clipping and gaussian stretch etc. The technical details of each enhancement procedure are as follows:

(i) *Linear Stretching*

It is a simple enhancement in which lower grew value "L" is mapped to a digital count 0 (zero) and the highest grey value "H" is mapped to a maximum digital count (usually 255) and the intermediate values are scaled proportionally, the two limits L and H are chosen from the input histogram (Jenson 1986).

(ii) *Linear Stretching with Clipping*

In this approach the input histogram is scanned through to determine the two limits for specifying the stretching function. This is done by specifying the percentage of the image that can be saturated (Moik, 1980).

(iii) *Gaussian Stretch*

The main characteristic of this technique is that the details within the extremes of the histogram can be preserved or increased by using cumulative Density Function of Gaussian distribution. And secondly there will be less emphasis in the Central (usually most populated) brightness zone (Jenson, 1986).

Study Area and Data Used

This study pertains to Tirupati town and environs which lies in Andhra Pradesh State. The latitude and longitude dimensions of the study area are 13′20 to 14′00 N/79 28 to 80 08 E. The details of data used is given in Table 22.1.

TABLE 22.1: Details of Data Sets Used for the Study

Sl.No.	Type of data	Data	Scale
1.	IRS-A LISS-1 BANDS 1, 2, 3 & 4 AND FCC ORIGINAL	18 MARCH 1988	1:250,000
2.	LINEARLY STRETCHED BAND 1, 2, 3 & 4 AND FCC	18 MARCH 1988	1:250,000
3.	LINEAR STRETCHING WITH CLIPPING BANDS 1, 2, 3 & 4 AND FCC	18 MARCH 1988	1:250,000
4.	GAUSSIAN STRETCH BANDS 1, 2, 3 & 4 AND FCC	18 MARCH 1988	1:250,000
5.	IRS-1A LISS-1 CCT	18 MARCH 1988	—
6.	TOPOSHEETS	—	1:250,000 & 1:50,000
7.	GROUND TRUTH	JANUARY 1988	—

Methodology

To evaluate the merits of each enhancement technique used here for identification and delineation of geomorphic features, a six-point grading assessment was followed as suggested by *IRS-1A Special Products Evaluation Committee, Report, 1988.*

A comparison of original image with various enhanced products, e.g. linear stretching, linear stretching with clipping and gaussian stretch products was made on the basis of interpretability of different geomorphic features, viz. flood plains, escarpment, Burried Pediment, Pediplain, Peneplain, ridge, butte and valley etc.

Results and Discussion

After the comparison of original image with different enhanced products a grading assessment was arrived at for identification of geomorphological features on the basis or interpretability which is as under:

Grading Assessment for Identification of Geomorphological Features on the Basis of Interpretability

Sl. No.	Feature	Original Image	Various Enhancements		
			Gaussian	Linclp	Lindir
1.	Flood Plain	3	5	6	4
2.	Escarpment	1	6	5	3
3.	Butte	1	6	5	4
4.	Peneplain	2	6	6	4
5.	Pediplain	2	6	5	4
6.	Buried Pediment	2	6	5	4
7.	Ridge	2	6	5	4
8.	Valley	1	6	5	4

6 Excellent
5 Very Good
4 Good
3 Distinguished (Dull Image)
2 Distinguished with Difficulty
1 Not Distinguished

A comparison of original image with various enhanced products, e.g. Linear stretching, Linear stretching with clipping and gaussian stretch shows that except flood plain other geomorphic features are not distinguishable in the original image. The observations comparing each enhanced product with original image are as follows – by using gaussian stretch escarpment, butte and peneplains are better delineated once the details within the extremes of histogram are increased by using CDF of Gaussian distribution. Linear stretching with clipping is very useful for identification and delineation of flood plains since 0.1 per cent of population is saturated after scanning through the input histogram on both sides of histogram thus features present on both the ends of histogram (water and dry sand) reveal better contrast preferably. For mapping of other geomorphic features, e.g. Burried pediments, ridges, pediplains and valley gaussian stretch is the best method.

Conclusion

A comparison of original image with various enhanced products shows that linear stretching with clipping is very useful for identification and delineation of flood plains. For the mapping of other landform features, e.g. Pediplains, ridges valleys, buried pediments, escarpments, buttes and peneplains etc. gaussian stretch is the best enhancement technique.

Acknowledgements

Authors are very grateful to Director, SAC, Ahmedabad for granting permission to present this research paper before the conference. The authors wish to thank Mrs Catherine Joseph for typing the manuscript neatly.

REFERENCES

Dhinwa, P.S., Jebakumar, M., Majumdar, K.L., Gopalan, AKS, Users' Handbook for IRS-1A Special Products, FRS/SAC/DP/TN-09/88 MAR-88, IRS-DP Group, SAC, Ahmedabad.

Jebakumar, M., Padmanabhan, K., Dhinwa, P.S., Gupta, P. Ramakrishnan, R. Majumdar, K.L. Mathematical Formulation and Detailed Design of Special Products Software, IRS/SAC/DP/TN-06/09, March 1986, IRS-DP Group SAC, Ahmedabad.

Jenson, J.R. (1986): *Introductory Digital Image Procession*, Prentice-Hall, Englewood Cliffs, New Jersey.

Moik, J.G., (1980): *Digital Processing of Remotely Sensed Images*, NASA, Washington DC.

23

Geomorphology of Rajasthan Through Remote Sensing for Environmental Management

SURENDRA SINGH, AMAL KAR and P.C. VATS

Introduction

UNTIL RECENTLY studies on the geomorphological aspects of India used to be carried out solely with conventional ground methods. Many of the studies, in spite of their excellent analytical interpretation of forms and processes, however, suffered from a serious drawback of correct demarcation of landform boundaries. For example, the delineation of boundaries between the flat units or those within an undulating terrain become very difficult and time-consuming with conventional ground methods and even impossible as the area of investigation becomes large. Experience has indicated that the best way to identify and map the different landforms is through the interpretation of remotely sensed data, especially the aerial photographs and satellite imagery, in conjunction with adequate ground verification survey. The synoptic data and the tonal differences provided by the different morphological situations, are major advantages over the field studies in quick and accurate mapping of the units, but for a correct judgement regarding the unit boundaries, especially in the flat areas, and for deciphering the reasons of tonal differences, detailed ground information is also required. For Rajasthan the significance of aerial photo-interpretation in the geomorphological investigations have been highlighted by several authors (Ghose and Singh, 1965; Ghose *et al.*, 1966; Singh and Ghose, 1969; Singh, 1983 and Ghose, *et al.*, 1983). Although low altitude conventional panchromatic aerial photographs are used widely for the purpose, use of recent satellite images, especially the US Landsat imagery, the Indian Remote Sensing Satellite images and the SPOT images have proved more useful in many areas, especially because of the different wavelength bands in which the images are registered by the Multispectral Scanners. For example, the near-infrared band (0.8-1.1 micrometers) is more useful for identification of moist zones, while the red band (0.6-0.7 micrometers) is useful for general distinction between the different landforms and delineation and mapping of their boundaries. Digitally enhanced satellite data have also been found very useful in detecting micro-variations within the landforms. It has helped in delineation of micro landform units with better precision than the standard false colour composites of the imagery (Kar, *et al.*, 1989).

Keeping in view the importance of remote-sensing techniques, investigations on the geomorphological aspects of Rajasthan State were carried out by using aerial and satellite remote sensing techniques.

Methodology

The study was conducted by using a three tier system viz., air, space and ground. Black and white vertical aerial photographs (1:30,000 and 1:25,000 scales), black and white and false colour composites of Landsat and IRS images in transparencies (1:1M scale) and in enlarged paper prints (1:50,000 scales and 1:250,000 scales): TERRA data (1:2,000,000 scale) and S.O.I. topographical sheets were used to conduct the geomorphological studies in different parts of Rajasthan. The preliminary boundaries of the landforms were delineated and mapped and field traverses were carried out to check the boundaries. Information on different morphological properties and associated hazards were also collected. Based on the laboratory and field data, final geomorphological map, showing the distribution of different landforms, was prepared for rational environmental management and planning.

Before discussing the results it is, however, necessary to provide some information on the evolutionary aspects of the landforms.

Landform Evolution

Rajasthan State, with its vast latitudinal (23°3′ to 30°3′N) and longitudinal (69°30′ E to 78°17′ E) extent, is composed of a variety of landforms which have evolved through the ages in response to the endogenic and the climatically controlled exogenic processes over various rock types. Geologically the State forms a part of the stable Indian (or the Peninsular) shield and the oldest rock formations like the pre-Aravalli granite-gneisses and a variety of quartzites, schists, gneisses, slates, marble along with intrusives of Archaean to Proterozoic era occur, especially in its eastern and central parts. These rocks, along with their associated tectonic elements suggest at least three cycles of orogenesis from the Archaean to the Proterozoic era, punctuated by igneous intrusions and long phases of sub-aerial denudation (Heron, 1953; Pareek, 1984 and Kar, 1983). The earliest known cycle, the Bhilwara cycle (more than 2500 m.y. old) started with deposition and geosynclinal development over the Pre-Archaean granite-gneiss basement. This was followed by the Aravalli cycle of orggenesis, marginally covering the Bhilwara geosyncline and having its axis roughly in NE-SW direction. A third cycle of orogenesis, the Delhi cycle, followed along the same axis and its culmination ended the long history of orogenesis within the State. The upheavals produced one of the oldest mountain ranges in the world (Heron, 1953 and Misra, 1967), the Aravallis, the surviving remnants of which still divide the State into major morphological units.

From the upper Proterozoic period onwards sedimentation started in several basins to the east and west of the Aravallis, under both continental and marine condition, producing the sedimentary beds of sandstone, limestone, shale, conglomerates, etc. The major igneous activities after the Aravalli orogenesis took place in the western part during the lower Palaeozoic as granite intrusions and rhyolite extrusions, now known as the Malani igneous suits and in the southeastern part during the cretaceous period as basalt flow, now known as the Deccan lava.

The present landforms have been sculpted over these various rock types through subaerial denudational processes in pre-Quaternary and Quaternary eras. The climatic variability within the State, from a subhumid southeastern part to the extremely arid western part, through a gradation of semi-arid and arid climatic regions, have also distinct control on the efficacy and distribution of processes. Thus, the fluvial erosional processes are more efficient in the eastern part, but gradually become less important in the western desert plains and negligible further west. The aeolian processes, on the other hand, are most vigorous along the extremely arid western and negligible from the Aravallis eastward. However, since the climate did not remain stable over the region, but fluctuated between drier and wetter phases, vestiges of the past climate-related features are also to be found in the seemingly unexpected locations (e.g. dissected sand dunes to the east of the Aravallis).

Discussion

Taking all these factors into consideration Rajasthan has been broadly divided into four major geomorphic regions. These are, from west to east, (1) the Rajasthan desert, (2) the Aravalli mountainous region, (3) the east Rajasthan plains and (4) the southeastern plateau region (Fig. 23.1). It is proposed to discuss the landform units in the non-desertic terrains on the basis of Landsat image interpretation, reconnaissance surveys and available secondary information, while for the desert information collected from field have been used in conjunction with the data from aerial photographs and satellite image interpretation, both visual and digital.

I. *Rajasthan Desert*

The desert in Rajasthan occurs to the West of the Aravalli mountain ranges. It merges westward with the desert in Pakistan, while in the south it is flanked by the Great Rann of Kutch. In the north the desert is flanked by fertile alluvial plains of Punjab-Haryana.

After the Aravalli orogenic cycle and associated igneous activities and the succeeding deposition of sedimentary beds of Marwar supergroup, much of the desert region, especially in its eastern part, experienced a long period of subaerial denudation. Deltaic to marine condition prevailed mainly in the western part, resulting in the formation of sedimentary beds of Permain to Tertiary in Bikaner-Jaisalmer-Barmer part. However, the morphological history could be reconstructed for the late quaternary period only, on the basis of available stratigraphic records. Studies have indicated that the desert was not as dry as it is now throughout the Quaternary period. At least two major wet phases prevailed here, followed by dry phases (Ghose, *et al.*, 1977a and Allchin, *et al.*, 1978).

The prolonged subaerial decomposition and denudation of the land surfaces produced extensive rocky/gravelly pediments through scarp retreat. These were then gradually covered with colluvial sediments by fluvial processes to form the flat buried pediments. At the distal end of the buried pediments, the extensive older alluvial plains were created during a major wet phase by the well integrated Aravalli and Himalayan drainage systems (Ghose *et al.*, 1977b; 1979). The illuviated caliche deposit (*Kankar* pan) within the alluvial sediments has been dated by C_{14} method as of 28000-30,000 year B.P. (Dhir, 1977) and the earliest age for the alluvial sediments is perhaps 40,000 years B.P.

This major wet phase was followed by a major dry phase, probably in pre-Holocene period and it ended around 10,000 years B.P. (Singh *et al.*; 1974). It was responsible for the formation of sand dunes of different types and sandy undulating buried pediments and plains (Singh and Ghose, 1977; Singh, 1981). As a result, most of the drainage channels were either buried under sand or were disorganised, creating closed inland basins. Another wet phase from 10,000 to 3800 years B.P. stabilised the active high dunes. The closed basins were filled with fresh surface run-off to form lakes. The second major dry phase set in probably around 3800 years B.P. but the climate slightly ameliorated around early century A.D. (Singh *et al.*, 1974). During this period, the fresh water closed basins turned into saline depressions, such as at Kaparda, Pachpadra, Malhar, Dediya, Bap and Didwana. The renewed aeolian activities created barchan and shrub-coppice dunes and extensive sand sheets in the western parts of the Rajasthan desert.

Thus the major existing landforms of the Rajasthan desert were created by the fluvial and aeolian processes. Mild tectonic activities also took place during the Quaternaries, affecting the fluvial activities (Kar, 1988, 1989 b).

On the basis of the above evolutionary history, form, pattern and characteristics of the drainage, slope and erosional/depositional/salinity hazards, Rajasthan desert has been divided into nine major landform units (Fig. 23.1). The distribution and salient geomorphological and Landsat spectral characteristics of these landform units are described below:

1. *Hills*: The hills composed of granites and rhyolites, occur in the form of domes and inselberge with tafonis. Around Sirohi, Abu and Erinpura, the hills are composed or Erinpura granite and pegmatite. The hills composed of sandstone and limestone of Marwar supergroup occur in the eastern parts of the Rajasthan desert, while the hills of

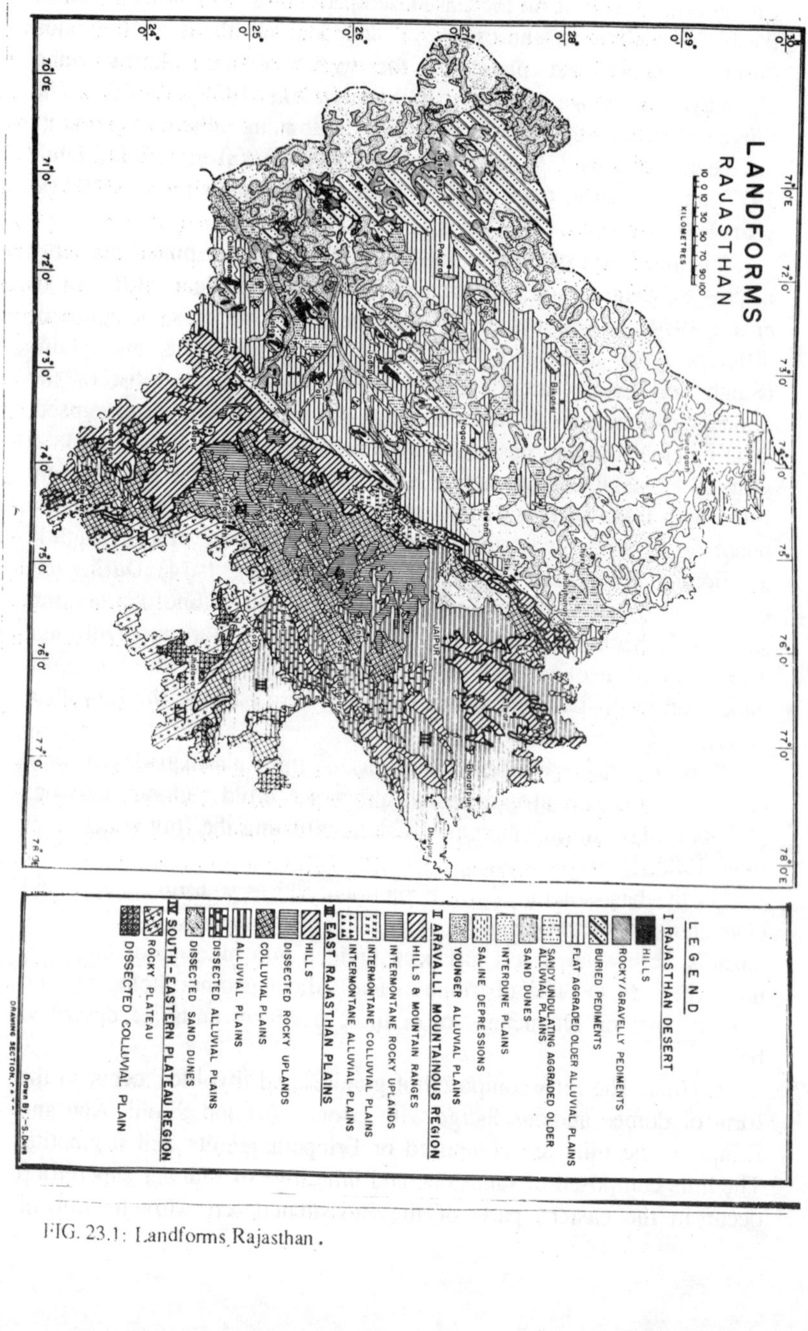

FIG. 23.1: Landforms Rajasthan.

later formations are found in Barmer, Jaisalmer and Bikaner districts. The sandstone hills are in the form of mesas and buttes with escarpment, while the limestone hills have subdued topography. The circular features of the igneous rocks which could not be detected from the field survey and interpretation of the aerial photographs are explicitly seen on the Landsat imageries (Plate 23.1).

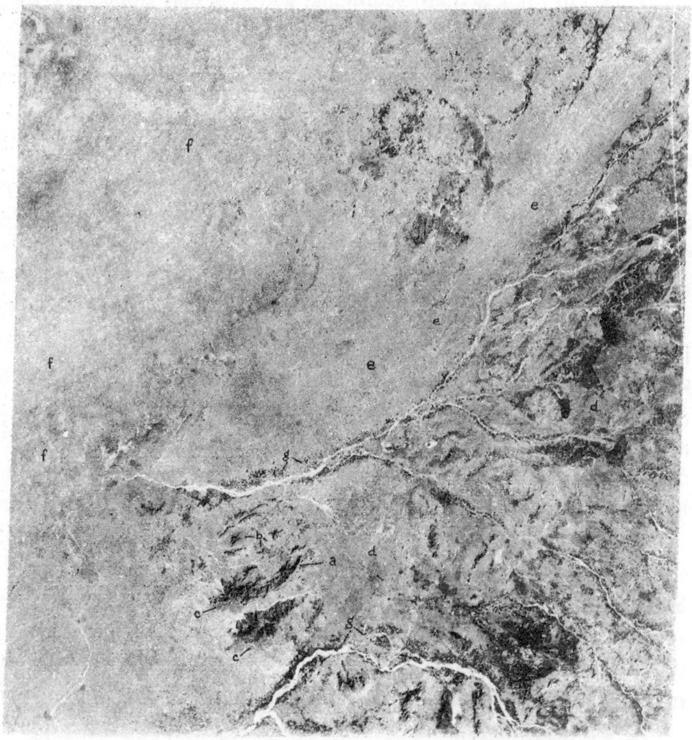

Plate 23.1: Landsat Imagery (Band 5) of Jodhpur — Jalor Area, Showing Spectral Characteristics of (a) Hills Around Siwana, (b) Rocky/Gravelly Pediments, (c) Dissected Sandy Undulating Buried Pediments, (d) Flat Older Alluvial Plains, (e) Sandy Undulating Older Alluvial Plains, (f) Stable Parabolic Dunes and Interdunal Plains, and (g) Younger Alluvial Plains.

2. *Rocky/gravelly pediments*: this unit occurs at the base of the hills and its slope varies from 3° to 8° in the upper part, 1° to 3° in the middle part and less than 1° in the lower part. The surfaces of these pediments are rocky excepting along joints where gravelly sediments have been deposited. The vast gravelly pediments around Kolayat and

Bap are the ideal examples of desert pavements.

This unit distinctly appears in light to dark grey tone in band 5 image and could be accurately separated from the hills and the buried pediments (Plate 23.1).

3. *Buried pediments*: Flat and sandy undulating buried pediments, covered with 25 to 40 cm deep sediments with less than 1° slope and 50 to 200 cm thick aeolian sediments and irregular slopes varying from 1° to 3° slope, occur at the flanks of the hills and rocky pediments. The alluvial and aeolian sediments are underlain by hard rocky strata. The diameter of the sediment particles varies from 0.06 to 0.85 mm. More than 65 per cent sediment particles are, however, of 0.06 to 0.12 mm size.

This unit distinctly stands out on the landsat bands 5 and 7 images in dark and white to dark grey tone. The dissections in the sandy deposits are clearly visible which helps in separating flat and sandy undulating buried pediments from each other (Plate 23.1).

4. *Flat older alluvial plains*: This unit is most extensive and occurs predominantly in the eastern part of the desert. These plains are covered with the alluvium of varying texture and thickness. In the southeastern part, the heavy textured alluvial plains at certain locations are saline. The depth of the *Kankar* pan varies from 7 to 180 cm. The nature of the sediments varies from loamy sand to sandy loam and also loam but at certain places in pockets silty clay loam are found. The size of the sediment particles varies from 0.06 to 0.84 mm; but more than 60 per cent of the sediment particles have diameter of 0.06 to 0.18 mm. Slope is less than 1° in this unit.

Light to medium and medium to heavy textured flat alluvial plains distinctly appear in grey to dark g.ey and dark tone on Landsat band 5 image (Plate 23.1). The saline plains exhibit white tone on this band.

5. *Sandy undulating older alluvial plains*: These plains are characterised by undulations in the form of sand sheets of 20 to 300 cm thickness, longitudinal and transverse dunes of 90 cm to 5 m height and sandy hummocks and ridges of 30 cm to 1 m height. The slope in this unit is irregular and varies from 1° to 3°. Within the sandy deposits, a weakly developed layer of $CaCO_3$ has been observed in some parts. The size of the sediment particles varies from 0.06 to 0.59 mm.

These plains, due to the sandy deposits and higher reflectance to green and infrared radiation, exhibit white to dark grey tone and distinctly appear on Landsat bands 5 and 7 images (Plate 23.1).

6. *Sand dunes*: Six types of dunes, belonging to the old and new dune systems, have been identified. The stabilised obstacle, parabolic, longitudinal and transverse dunes of 10 to 80 m height belong to the old dune system. The active shrub-coppice and barchan dunes of 90 cm to 2 m and 3 to 10 m heights respectively are of new dune system. The size of sand grains varies from 0.06 to 0.25 mm. The average diameter of the sand grains ranges from 0.11 to 0.12 mm.

The sand dunes of different types clearly stand out of Landsat band 7 imagery in white tone due to the higher reflectance from the vegetation (Plates 23.1 and 23.2). These could be accurately separated and mapped from the interdune plains which appear in dark to dark grey and light tone.

7. *Interdune plains*: These plains are flat and sandy undulating and occur between the stabilised dunes. The slope of the flat interdune plains is less the 1°, while that of sandy undulating interdune plains is irregular, varying from 1° to 3°. The sediments of these plains are alluvial and aeolian and the size of these sediment particles varies from 0.06 to 1.19 mm, but 60 per cent of the sediments are of 0.06 to 0.12 mm size.

The Landsat spectral characteristics of this unit are similar to the flat and sandy undulating alluvial plains (Plates 23.1 and 23.2).

8. *Shallow saline depressions*: Large and small saline depressions of different shapes and sizes occur at Sambhar, Kuchaman, Didwana, Lunkaransar, Jamsar, Bap, Thob, Pachpadra, Sanwarla and around Pokaran and Jaisalmer. Among these, the Sambhar saline lake covers the largest area. (Ghose, *et al.*, 1977b). The size of sediment particles varies from 0.06 to 1.19 mm and the particles of 0.06 to 0.07 mm diameter are predominant.

These depressions, due to their shape, size and location, appear distinctly in white to dark grey tone on all the four Landsat bands (Plate 23.3).

9. *Younger alluvial plains and river beds*: These plains occur in narrow strips along the courses of the Luni and the Ghaggar rivers and their tributaries (Plates 23.1 and 23.4). The plains constitute deep to very sediments varying from sand to loamy sand and sandy loam, while the river beds comprise gravel and coarse sand. The size of the sediment particles varies from 0.06 to 1.19 mm and the mean diameter of the particles is 0.39 mm. In some parts, especially near the banks, sandy hummocks of 50 cm to 2 m height and longitudinal and transverse dunes of 3 to 6 m height have created the undulations.

Plate 23.2: Landsat Imagery (Band 7) of Area to the North-west of Jaisalmer, Showing (a) Longitudinal and (b) Transverse Dune Fields with Interdunal Plains.

These plains along the river beds, due to their location and linear pattern, are distinctly visible in dark to dark grey and white tone on Landsat imagery (Plate 23.1).

Assessment

The younger alluvial plains and the flat buried pediments, having level slope, better soils and groundwater conditions and absence of hazards, are the most suitable resource units for the development of irrigated farming. In the flat older alluvial plains, some parts due to the presence of the buried channels are the potential sites for groundwater exploration and development of irrigated farming. The large acreage of these plains, due to limited hazards, level slope, better sediment and moisture conditions, are best suited for dry land agriculture.

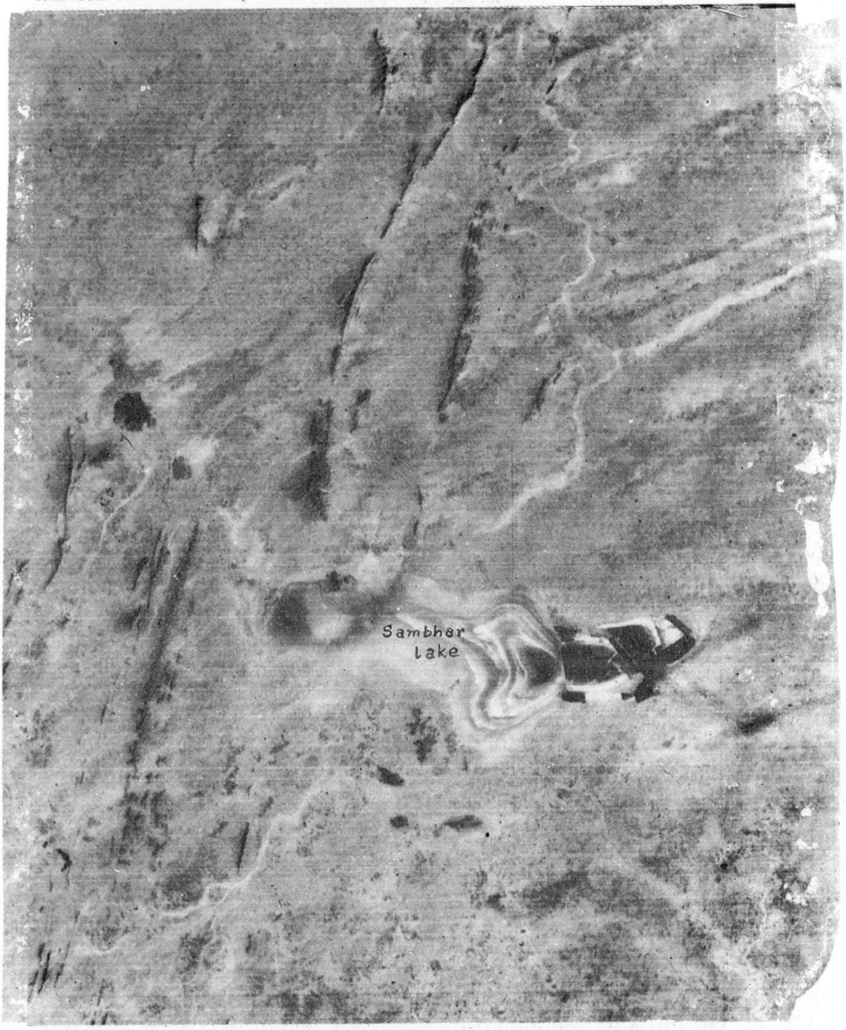

Plate 23.3: Landsat Imagery (Band 5) of Sambhar Saline Depression.

Sandy undulating alluvial plains and interdune plains, due to severe to very severe wind erosion/deposition hazards and poor fertility status of soils, should be occasionally cultivated by using suitable soil conservation measures such as wind breaks and wind strip cropping. The sand-dunes, which have several physical and climatic limitations such as undulating topography, irregular slope, deep to very deep sandy soils with low fertility status, wind erosion/deposition hazards, poor rainfall and high temperature regime, could be developed for silvipastures, Over-exploitations in many of the units are presently leading to wasteland formation (Singh & Kar, 1988).

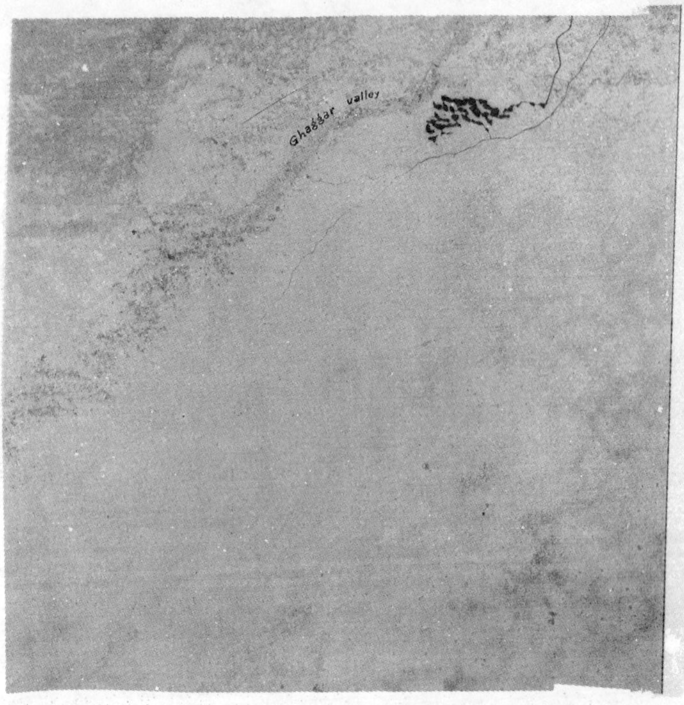

Plate 23.4: Landsat Imagery (Band 7) of the Ghaggar Younger Alluvial Plain in Ganganagar District of Rajasthan.

II. *Aravalli Mountainous Region*

The Aravalli hills and mountainous region, one of the oldest surviving mountainous belts in the world, divides the State into western and eastern zones. Most of the major rivers within the State originate from the Aravallis (Fig. 23.2) and a number of these are structurally controlled. The only exceptions are the Chambal and its tributaries, the Kali Sindh and the Parbati, which originate from the plateau region to the southeast (Fig. 23.2). Because of long period of denudation the hill ranges do not run in continuous chains, but are broken at several places, especially in the central part. The main ranges are generally formed of quartzite and granite, while the schists and gneisses form the rolling uplands and broad valleys. After the Proterozoic orogenesis the Aravallis experienced long phases of subaerial denudation, producing low hill ranges and monadnocks, interspersed with rolling uplands. From accordant summit levels Heron (1953) identified three peneplains in the Aravallis and the adjoining eastern plains and termed them as the Mesozoic, Tertiary and Pleistocence peneplains. Upliftment took place along the Great Boundary Faults flanking the then rolling hilly topography after each of these peneplanation to such a way that during each movement the southern part of the mountainous region was upthrown more compared to the northern part, Thus giving rise to the present distribution of maximum number of high peaks in the southern part. From the characteristics of the hill ranges and associated plains the region could be broadly divided into (a) the southern Aravallis, with high relative relief, closely spaced hill ranges and high peaks; (b) the central Aravallis with wide valleys and low hill ranges (Plate 23.5) and (c) the northern Aravallis, with relative relief higher than in the central part and few high peaks.

The following are the major landform units (Fig. 23.1) within the Aravallis:

1. *Hills and mountain ranges*: Most of the hill ranges are composed of rocks of Delhi system such as calc-schist, calc-gneiss, biotite-schist, phyllite, crystalline limestone, quartzite, etc., although the rocks of Aravalli system do also form them, especially in the southeast, roughly between Kankroli and Dungarpur. The most resistant hills with higher relief are those formed of quartzite, e.g., near Alwar, Ajmer, Gogunda.

In this part, due to better rainfall as compared to the western part, the hill slopes are generally debris-covered and support good vegetation. In fact, as one approaches the Aravallis from the western desert plains and moves within, the influence of slightly better climate is felt through the increased production of debris along the hill slopes, irrespective of rock type, and their part retention on the steep slopes. The occurrence of a talus slope at the foot hills, the length of which increases as one proceeds eastward, and summit rounding are the other significant characteristics of the hills within the region (Kar, 1983b). The situation is slightly different where granites form the hill ranges, e.g. along the southwestern margin where the Erinpura granite and the Sendra granite are the chief constituents. In these cases the hill slope is generally barren, without appreciable debris cover, have convex outline and are susceptible to chemical weathering, producing cavernous features of various shapes and sizes and spheroidal boulders (Kar, 1983 b).

Most of the westernmost hill ranges to the north of Bar are affected by aeolian activities in the form of obstacle dunes which are also highly dissected by rills and gullies. Although such dunes are mainly concentrated along the western slopes of the outer hill ranges, in some cases they crossed over to the other side to form leaward obstacle dunes.

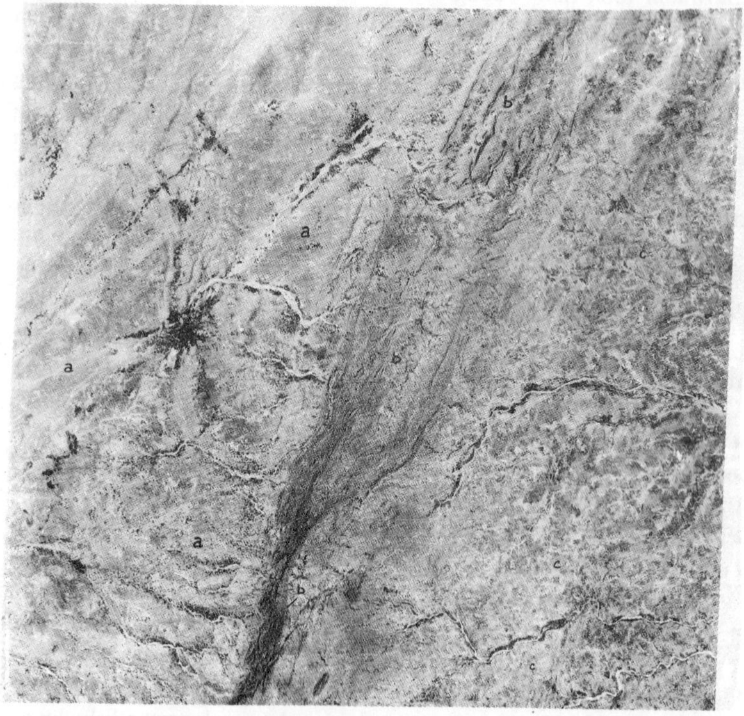

Plate 23.5: Landsat Imagery (Band 5) of Bilara – Ajmer – Bhilwara Area, Showing (a) The Eastern Part of the Desert Plains, (b) The Central Part of the Aravallis and (c) The Dissected Rocky Uplands with Shallow Skeletal soils in the East Rajasthan Plains.

FIG. 23.2: Drainage Rajasthan.

2. *Intermontane rocky uplands*: In many parts, the hill ranges are flanked by undulating rocky uplands dotted with low hills. The topography develops especially in areas of granite-gneiss, phyllite, biotite-schist and slate. The unit is characterised by rocky/gravelly surface and is dissected by numerous rills and gullies. The slope is irregular and varies, on an average, from 1° to 5°, but is more in case of low hills on them.

3. *Intermontane colluvial plains*: These plains with 0° to 1° slope have been created by fluvial depositions, especially through sheet floods and activities of mountain streams. The plains are formed of heterogeneous mixture of coarse sand, gravels and pebbles along with finer particles. The thickness generally varies from 45 cm to 1m. To the north of Ajmer the sediments are partly mixed with aeolian sand, the genesis of which could be traced to the earlier dry phase when the desert extended beyond the Aravallis. The colluvial plains are the widest in the central part where the hill ranges are wide apart. Some

areas of sandy undulating colluvial plains have also been noticed within this unit near Pushkar.

4. *Intermontane alluvial plains*: These alluvial plains have been formed due to the flooding and alluviation along some of the major streams such as the Saraswari and the Sagarmati. The depth of alluvium is more than 3 m and the sediment texture varies from sand to loamy sand, underlain by the riverine pebbles and then hard rock at a greater depth. The slope is less than 1°. This unit, along with the intermontane colluvial plains constitute the best agricultural lands within the Aravallis, because of flat terrain and proximity of groundwater.

Assessment

Compared to the desert region the Aravallis lie in a better climatic belt, with annual rainfall of 500 mm to 1000 mm. As a result there is better soil cover, even along the hill slopes and higher drainage network. The region should have, therefore, a better outlook from resource potential and utilisation points of view. However, since the climate is semi-arid, slope is relatively steeper and whatever thickness of sediments is there develops very slowly, the indiscriminate and unmanaged felling of trees and cultivation along the marginal lands have accelerated the soil erosion to such an extant that many of the hills are becoming barren, rocky and highly dissected by rill and gullies. The Aravallis have, therefore, become one of the most threatened regions within the state and the problem is increasing at an alarming pace in the hills and in the foothills regions. In the areas where obstacle dunes form a part of the foothills, the devegetation of the slopes are creating localised problem of rapid development of gully networks. In the Pushkar-Budha Pushkar area over-exploitation of the stabilised fossil dunes and the sandy intermontane plains have created localised problems of reactivated dunes and new low dunes on the fertile plains. Such trends could, however, be controlled through careful plantation programme in the area (Kar, 1986). However, the colluvial and the alluvial plains remain the best agricultural units within the region, because of better groundwater potential and flat terrain.

III. *East Rajasthan Plains Region*

To the east of the Aravalli mountainous region and up to the Chambal valley, the east Rajasthan plains region is a vast plain formed on gneisses and schistose rocks of the Bhilwara and the Aravalli Supergroups. In the north-east it merges with the Yamuna plains, while in the southeast it abruptly ends at the margin of the south-eastern plateau region. At least two erosional surfaces could be identified within this region. The older and the most widespread surface is represented by the shallow to moderately deep plains formed on gneisses and schistose rocks and could be the Tertiary peneplain of Heron (1953). The younger one is being formed presently through vigorous downcutting and plantation processes in the recent alluvium along the Chambal and its tributaries. Both of these could be genetically related to the tectonic movements along the Great Boundary Fault during the Himalayan orogenesis in the Tertiary period and during the Pleistocene to recent period.

The major landform units within this region (Fig. 23.1) are as follows:

1. *Hills and hill ranges*: Excepting the hills in the southeastern part, especially to the east of a line joining Bundi and Chittorgarh, most other hills are formed of quartzite and granite of the Bhilwara cycle. They have conical to rounded shape and could be taken as monadnocks over the vast peneplane surface. To the east of Chittor-Bundi line the hills are formed of sandstone and have flat summit. The most prominent hill ranges occur between Bundi-Sawai Madhopur-Bayana, roughly along the Great Boundary Fault. Another series of dissected low hills occur in a linear fashion from the west of Bhilwara to Toda Bhim, through Jahazpur, Tonk and Lalsot. Like the hills in the Aravallis, most of these hills have appreciable colluvial debris.

2. *Dissected rocky uplands with shallow skeletal soils*: This unit, formed mainly on the gneissic and schistose rocks, occurs along the Aravallis, roughly up to Bhilwara and Tonk. The unit is characterised by broad undulations with rocky exposures or very shallow soil cover (Plate 23.5). Numerous smaller tributaries of the Banaes, the Berach and the Khari have dissected this upland through vigorous headward erosion, leaving narrower interfluves.

3. *Colluvial plains*: The dissected uplands are flanked to their east by shallow to moderately deep colluvial plains, especially along the major tributaries of the Banas. As in the case of intermontane colluvial plains, this unit is also formed of heterogeneous mixture of gravels, coarse sand and finer particles and the thickness generally varies from 50 to cm to 1 m.

4. *Alluvial plains*: The major streams such as the Chambal, the Banas, the Bandi and the Banganga have built up their own alluvial plains to the east of the colluvial plains unit. The unit is more widespread and thicker in Jaipur-Bharatpur area, where the alluvium is mixed with aeolian sand of various thickness.

5. *Dissected alluvial plains*: Along the Chambal and its tributaries such as the Parbati, the Kali Sindh, the Kural, the Mej and the Sip, and along some tributaries of the Banas, such as the Morel and the Dhil, the alluvium has been badly dissected through headward erosion of gullies, so much so that the topography has been changed from an original flat alluvial plain to a ravine-infested 'badlands' (Plate 23.6). The genesis of this badland topography could be traced to a recent change of base level through tectonic movement along the Great Boundary Fault. Sharma (1968, 1979, 1980) has dealt with in details the various factors responsible for the evolution of the badlands (ravines).

6. *Dissected sand dunes*: Between the Sambhar lake and Jaipur considerable area is under 5 to 15 m high sand dunes. The dunes were formed in the earlier dry phase when the desert extended beyond the Aravallis (Ghose, *et al.*, 1977a). All these dunes are stable, well vegetated and dissected by rills and gullies. They differ from the dunes of the desert proper in not having reactivated crests and flanks and in the degree of compaction dissection. These dunes are completely stabilised and are in eluvial phase where illuviation process is dominant.

Assessment

Climatically the east Rajasthan plains region lies in an advantageous position than the plains in the west. The watter climate helps in increasing the potentiality of the vast area under uplands with shallow soils and the other colluvial plains along tributary valleys. Almost similar heterogeneous mixture of sediments with identical thickness in the desert have much less potentiality because of the much restricted soil forming processes under lower rainfall condition. The alluvial plains in the northeast and along the major streams elsewhere within this region are the most favourable locations for agricultural purposes, because of much better soil depth and good quality groundwater at shallow depth. However, in spite of the region being located in a climatically advantageous position, some of the landforms are highly vulnerable to erosional processes. The dissected dunes of

Sambhar-Jaipur-Kishangarh-Ajmer region are widely cultivated, but near the hills and in other gullyprone areas these dunes should be planted with trees, so that soil erosion is minimised. Similarly afforestation in the ravine-infested areas will reduce the pace of erosion. It is suspected that devegetation plays a significant role in the accelerated erosion in both the above units.

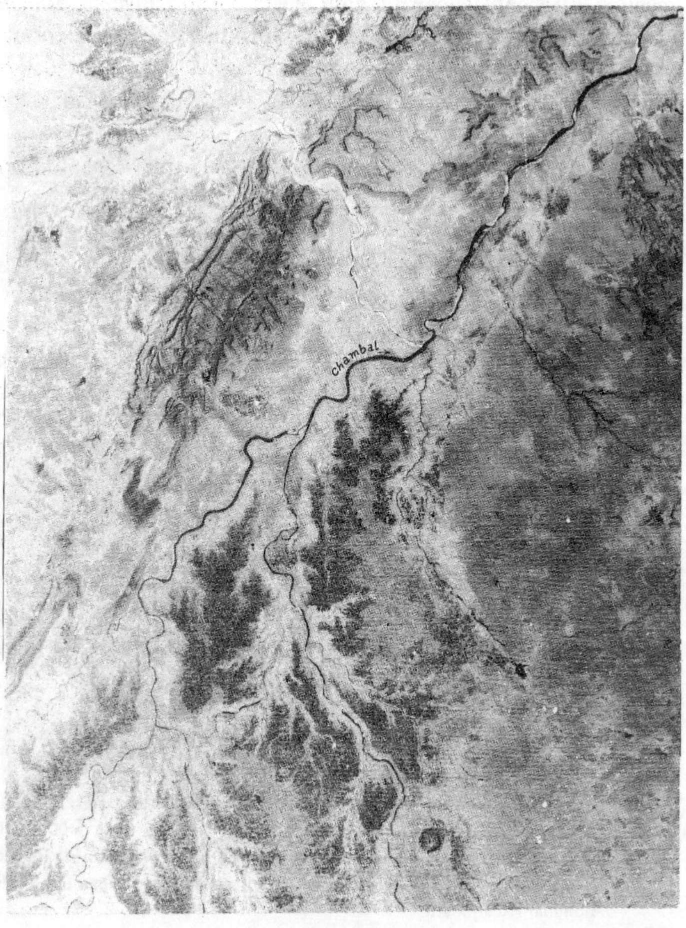

Plate 23.6: Landsat Imagery (Band 5) of the Chambal Ravines – The Dissected Alluvial Plains Along the Chambal and its Tributaries.

IV. *South-eastern Plateau Region*

The extreme southeastern part of the state is a plateau capped by the basaltic beds of Deccan Lava. It rises sharply from the adjoining east Rajasthan plains along an escarpment (Plate 23.7). Although basalt generally forms the capping, Vindhyan Sandstone sequences have also been exposed over large areas due to erosion by the Chambal river system. The Chambal emerges from this plateau through a deep gorge, suggesting a recent change of base level. This change in base level might also be the cause of vigorous downcutting and ravine formation in the plains below.

Plate 23.7: Landsat Imagery (Band 5) of Chittaurgarh — Kota Area and the Upper Catchment of the Chambal Showing (a) The Shallow Colluvial Plains with Rock Outcrops in Kota Area and (b) The Southeastern Plateau Region Rising Along an Escarpment.

Two major landform units could be identified through Landsat image interpretation of the area (Fig. 23.1).

1. *The rocky plateau*: This unit is widespread and occurs mostly as a flat surface dotted with steplike formation of hills with flat summit and steep escarpments. The Vindhyans and the Deccan Trap formation form the rocky exposures and parent material for colluvium everywhere.

2. *The dissected colluvial plain*: This unit occurs along the major streams and their tributaries. The surface is seldom flat and undissected because of high drainage density (Plate 23.7). From the imagery the unit resembles the dissected younger alluvial plains further downstream. However, sediment depth is much shallower here and has a heterogeneous mixture of finer and coarser fragments.

Assessment

The colluvial plains are the best areas for agricultural activities and are widely used for the purpose. However, the higher drainage density puts a limit to cultivable area and needs suitable measures such as contour bunding for soil and water conservation. The rocky plateau, due to physical limitations such as rocky surfaces and lack of soil depth is suitable for afforestation. Soil conservation measures and afforestations are to be continued in this part for restricting the soil loss through gully formation and to save the two major dams on the Chambal from siltation.

Conclusion

Rajasthan State, based on its present morphological set-up and the morpho-climatic processes operating on them, has been divided into four major geomorphic regions. Each of them has again been subdivided into broad landform units. Thus, the Rajasthan desert has nine major landform units, the Aravalli region has four, the east Rajasthan plains has six and the plateau region has two. Land being central to any natural resources development planning and in the assessment and management of natural environment, such geomorphological classification and analysis plays a key role in correct identification and assessment of the resources, as has been indicated in this article. For example, the genesis of ravinelands along the Chambal could be linked with rejuvenation and man-induced accelerated erosion. The accelerated soil erosion could be substantially reduced through soil conservation measures and afforestation, but the vigorous natural process of stream's adjustments to the new base level could hardly be stopped through such activities. Within the desert most of the aeolian landforms are highly vulnerable and hence any programme like sand dune stabilisation or shelter belt plantation should be based on proper information on the genesis and morphology of the aeolian landforms and the expected reaction of it to any such human activity. Similarly the stabilised fossil dunes within the Aravallis and in the eastern plains need not necessarily be thought as dangerous sites for cultivation. At very few localities extreme over-exploitation of such dunes may have created localised problems of reactivation and new dune formation, but the situation could be reversed through careful planning at those areas (Kar, 1986; 1989b). It does not indicate an invasion of the desert proper to the problem site. All the geomorphic regions have their own problems and potentialities, based on their typical landform assemblages and other morphological characteristics. Therefore, any programme for improvement of the natural environment in these regions should be based on those geomorphological characteristics.

Acknowledgements

The authors wish to express their sincere thanks to Dr. K.A. Shankarnarayan, Director and Dr. R.P. Dhir, Head of Basic Resources Studies Division, Central Arid Zone Research Institute, Jodhpur, for their encouragement and the facilities provided to write this article.

REFERENCES

Allchin, B., Goudie, A. and Hegde, K.T.M. (1978): *The Prehistory and Palaeogeography of the Great Indian Desert.* Academic Press, London.

Dhir, R.P. (1977): Palaeoclimatic inferences from Quaternary pedogenetic processes in arid zone soils. In *Ecology and Archaeology of Western India* (Ed. D.P. Agrawal and B.M. Pandey). Concept Publishing Co., New Delhi, pp. 127-134.

Ghose, B., Kar, Amal and Zahid Husain (1979): The lost Courses of the Saraswati river in the Great Indian Desert - New evidence from Landsat imagery. *The Geographical Journal*, 145 (3): 446-451.

Ghose, B., Pandey, S and Singh, S. (1966): Studies in Photogrammetry. Geomorphic studies of Sojat from aerial photographs to delineate landforms. *J.Ind.Soc. Soil Sci.* 14: 9-12.

Ghose, B., and Singh, S. (1965): Geomorphological studies in Rajasthan I. Analysis of Saila block from aerial photographs. *Ann. Arid zone 4* (1): 49-55.

Ghose, B., Singh, S. and Kar, Amal (1977a): Desertification around the Thar - A geomorphological interpretation. *Ann Arid Zone* 16(3): 290-301.

Ghose, B., Singh, S. and Kar, Amal (1977b): Geomorphology of the Rajasthan desert. In, *Desertification and its Control* (Ed. P.L. Jaiswal *et al.*), I.C.A.R., Model Press (P.) Ltd. New Delhi, pp. 42-57.

Heron, A.M. (1953): *The geology of central Rajasthan*. Memoirs of the Geological Survey of India, *79*; 389 p.

Kar, Amal, (1983a): Stratigraphic sequence and geological history of the upper Luni basin. In, Shankarnarayan, K.A. and Kar, Amal (ed.), *Upper Luni basin - an integrated analysis of natural and human resources for development planning*. CAZRI, Jodhpur; 24-41.

Kar Amal (1983b): Morphology and evolution of slpes in the upper Luni basin, Rajasthan. *Annals of Arid Zone*, 22(4): 367-375.

Kar, Amal (1986): Physical environment, human influences and desertification in Pushkar-Budha Pushkar lake region of Rajasthan, India. *The Environmentalist*, 6 (3): 227-232.

Kar, Amal (1988): Possible neotectonic activities in the Luni-Jawai plains, Rajasthan. *J. Geol. Sec. India*, 32 : 522-526.

Kar, Amal (1989a): Evidence for neotectonism from the Indian desert. In, Singh, S. and Tiwari, R.C. (eds.), *Geomorphology and Environment*. Allahabad Geogr. Soc., 300-310.

Kar, Amal (1989b): Evaluation of present-day geomorphological processes in the Indian desert for assessing land degradation and natural hazards. *Ibid.* 643-654.

Kar, Amal., Singh, S., Vangani, N.S., Dubey, J.C. and Ghosh, T.K. (1989): Application potentials of IRS-IA data for mapping hydrogeomorphic features in the Indian desert. *Ext. Abst., Regional Workshop Ind. Rem. Sens, Sat. Mission and Its Application Potential*, RRSSC, Jodhpur, 54-58.

Misra, V.C. (1967): *Geography of Rajasthan*, National Book Trust, New Delhi; 188 p.

Pareek, H.S. (1984): *Pre-Quarternary Geology and Mineral Resources of North-Western Rajasthan*. Geol. Surv. India, Mem, 115; 99p.

Sharma, H.S. (1968): Genesis of ravines of the lower Chambal valley, India. *Proc. 21st International Geographical Congress, India*, Vol. 1: 114-118.

Sharma, H.S. (1979): *Physiography of the lower Chambal valley and its agricultural development*. Concept, Delhi.

Sharma, H.S. (1980): *Ravine Erosion in India*. Concept, Delhi. 96p.

Singh, G., Joshi, R.D., Chopra, S.K. and Singh A.B. (1974): Late Quaternary history of vegetation and climate of the Rajasthan desert. *Phil. Trans. Royal Soci. London*. 267 (b) : 467-501.

Singh, S. (1981): Types and formation of sand dunes in the Rajasthan desert. *In. Perspective in Geomorphology* (Ed. H.S. Sharma), Concept Publishing Co. New Delhi, Vol. 4, pp. 165-183.

Singh, S. (1983): Application of remote sensing techniques in geomorphological investigation - A case study of Jodhpur district, western Rajasthan, India, *Ann. Arid Zone*, 22 (4) : 283-294.

Singh, S. and Ghose, B. (1977): Geomorphology of Luni basin and its palaeoclimatic inferences. *In, Ecology and Archaeology of Western India* (ed. D.P. Agrawal and B.M. Pandey. Concept Publishing Co. New Delhi. pp. 136-148.

Singh, S. and Kaith, D.S. 1971): Geomorphological analysis of landform features of Pali development block from aerial photographs. *Ind. J. Geog*., 46 (3 & 4) : 50-59.

Singh, S. and Shankarnarayan, K.A. (1982): Application of Landsat digital analysis in geomorphological investigations of the middle Luni basin, Rajasthan, India, *Ann. Arid Zone*, 21 (4) : 279-287.

Singh, S. and Shankarnarayan, K.A. (1985): Remote sensing in monitoring natural resources and environmental hazards in the Indian desert. *Proc. Sixth Asian Conf. Remot. Sens. Hyderabad. India* : 35-41.

Singh, S., Vats, P.C. and Kaith, D.A. (1982): Some aspects of desert geomorphology of Bikaner district, western Rajasthan. *Geol. Surv. India, Miscellaneous*, No. 49 : 29-36.

Singh, S. and Kar. Amal. (1988): Geomorphological appraisal of the arid wastelands by using remote sensing techniques. In, Shankarnarayan, K.A. (ed.); *Wastelands Development and Their Utilization*, Scientific Publ., Jodhpur; 62-86.

24

Landscape Analysis of Desertification in Rajasthan by Satellite Imagery

ALEXANDER S. SHESTAKOV

RECENTLY A complex of problems connected with rapid rate of desertification in a number of the planet regions, including subhumid tropics became more acute. Three tendencies are clearly traced in development of natural resources in arid areas. They are: (1) growth of arable lands which is firstly due to ploughing up of marginal and unproductive lands, (2) decreasing of the pasture areas with simultaneous rising of animal population, (3) general elimination of tree-shrub vegetation for the purposes of fuel, building materials, food and fodder. Consideration of land resources and control of their condition has become of great importance for objective evaluation of the character of anthropogenic influence on arid environment. To solve these questions satellite imageries can be used on a wide spectrum of application.[4,6,8,9] Satellite imagery (SI) contains information about a number of interconnected environmental components. This fact allows to consider all natural objects simultaneously and in complex during the process of investigation and mapping of the territory. Moreover, the materials of space survey give the more precise and objective idea about peculiarities of the territory and boundaries of the objects. While investigating the desertification with the help of SI it is necessary to take into account the specific features of the landscape so optical characteristics of some of their components can reflect certain alterations of the environmental conditions. Thus, different spectrum brightnesses will characterised disperse and degraded vegetation, concentration of salts in upper soil layer, increasing of soil water capacity and some other features and processes in landscapes.[6]

In India, where arid and semi-arid lands occur about 320-000 sq km the north-western districts are to the greatest extent subjected to desertification.[16] These areas are famous with their very high population density (from 4 to 157 persons/sq km with the average 40 persons/sq km) in comparison with other arid regions of the world (3 persons/sq km).[12] Some negative adverse side effects of environmental degradation are widespread under the conditions of intensive anthropogenic pressure in Rajasthan. In connection with studying of desertification in this region, a landscape analysis of a part of Rajasthan plains has been made. The task of our investigation was to determine the leading factors and forms of desertification, their interconnection and dependence on landscape characteristics. All work is based on results of satellite imagery deciphering.

Materials and Methods

The Landsat false colour satellite imagery of a part of Rajasthan plains near the central part of the Aravalli range (about 30,000 sq km) has been used. The SI was made in December 1972 in 1:1,000,000 scale.[14] A visual method of deciphering was used according to basic indicators such as outline of imagery, tone, shape and size contours, colour, boundary characteristics etc.[5] Maps of India were used for the proper landscape regionalisation.[1,13,19] Landscape analysis was carried by distinguishing natural-territorial complexes.[15]

Results and Discussions

Examining SI embraces a number of landscapes from arid in the west to subhumid in the south-east. Common features of climate which determine conditions of desertification development are typical for the whole area, for example: high average annual temperatures, extreme unevenness of precipitation within the year, often recurrence of droughts, great number of sand and dust storms, severe wind from arid regions. During June-September usually about 80 per cent of annual precipitation falls.[18] Geomorphological, hydrological and soil factors become more important as natural prerequisites of desertification in conditions of comparatively similar climatic background with common tendency for decreasing of rainfall from 600-700 mm in the south-east to 300 mm and less in the west and increasing of aridity at the same direction.

We marked out 20 landscape complexes within 8 big landforms on the studying territory (Fig. 24.1). In the east old crystalline basement of Indian platform is covered by thin mantle of loamy unconsolidated sediments. In this part relief is represented by socle plains. Denudation plateau forms in the areas of basement outcrops. In the west a border part of basement massive is diped and covered by the thickest alluvial sediments which are represented on the surface by old alluvium of light composition and recent aeolian sandy deposits. There is a wide spectrum of soils from red-brown in the east to light-brown and sernozems in the west. Presence of light soils in the west is a condition for intensive development of wind erosion and formation of aeolian relief, for growth of sand fraction in humus horizon of soils. Heavy soils in the east are prone to salinisation. Two zones of plains are divided by the massive of one of the oldest retaining in the Earth relief mountains Aravalli. The mountains formed by the complex of crystalline rocks rapidly turn into bold outcrops with elimination of vegetation.

Seasonal prevalence of precipitation and their deficiency determine poor development of watercourse network. The Aravalli is the main watershed and centre of rivers origin for this territory. On the east where depth of incic and water filtration are limited with downpour regime of precipitation floods become disastrous. Linear and especially sheet erosion have widespread development. Gully, intricate and ramified erosion network is clearly traced on SI by specific drawing and light tone of imagery (Fig. 24.3). On the west where temporary streams prevail in rainy season they bring great quantity of detrital material and sand. During dry period along numerous dry rivers beds these materials become loose and undergo wind erosion resulting in aeolian forms. Unconfined groundwater is almost absent on platform part of region but they play an important role for development of irrigation on the old alluvial plains. However, these waters are characterised by rather deep occurrence and often high mineralisation which grow towards west. On SI the areas of irrigated land on the west are clearly registered by piedmont plain zone and only some separate plots can be find along river beds. Wide use of saline water for irrigation can lead to uprising of underground and subsurface water levels and then to secondary salinisation.

Sparse natural vegetation is represented by xeromorphic deciduous forests on mountain slopes in more humid areas and by torn open woodlands, shrub and deserted savanna, tree-and-shrub semi-desert on other surfaces.[3,17] According to the common tendency aridity of the

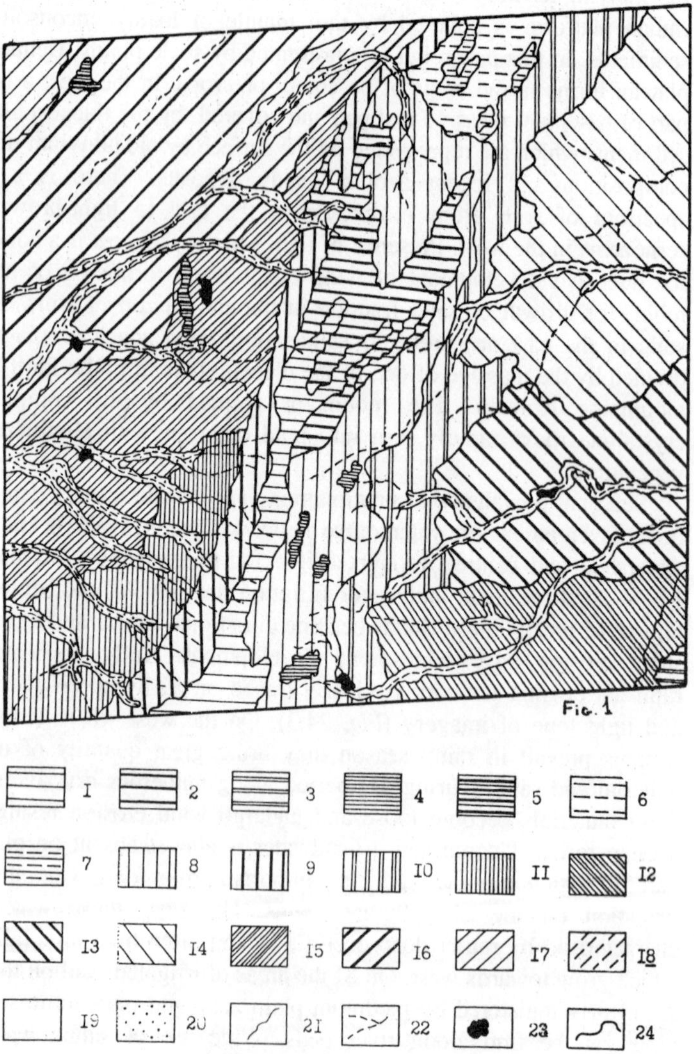

FIG. 24.1: Landscape structure of central Rajasthan (central part of Aravalli range). I. Folded block-mountains formed by complex of gneiss, metamorphic schists, granite: 1 — middle mountains with xerophitic desiduous forests on primitive rubbish red-brown soils; 2 — low mountains with open woodland and shrubs

on rubbish red-brown soils. II. Degraded denudation monoclinal plateau formed by Vindhayan sandstone and schists with deciduous forest on red-brown soils. III. Residual-mountains and mountain ridges: 4 — formed by metamorphic quartzite with xerophitic deciduous forests on primitive rubbish red-brown soils; 5 — formed by sandstone with torn woodlands on rubbish red-brown soils IV. Peneplain formed by alluvial-deluvial sediments with residual mountain-ridges: 6 — high alluvial-denudation plain with deserted savanna on grey-brown soils; 7 — residual mountain ridges formed by schist, granite, sandstone with open woodlands on rubbish red-brown soils. V. Piedmont plains: 8 — erosion-denudation pediments with open'woodlands on red-brown soils; 9 — erosion-deflation pediments with open woodlands and shrub savanna on serozems; 10 — colluvial-alluvial with open woodlands on loam red-brown soils; 11 — alluvial and alluvial fan deposits with srub savanna on sandyloam grey-brown soils. VI. Hillock erosion socle plains: 12 — with xerophitic decidious forest on loam red-brown soils; 13 — with open woodlands and shrub savanna on loam red-brown soils; 14 — with deserted savanna on sandy-loam red-brown soils. VII. Old alluvial plains: 15 — flat accumulative-denudation with deserted savanna on sandy-loam tropical grey-brown soils; 16 — flat inclined accumulative-deflation with deserted savanna on sandy-loam grey soils with *kankar*; 17 — sloping hillock sand accumulative with deserted savanna on sandy tropical light-brown soils; 18 — flat inclined aeolian with undershrub semidesert on sandy grey-brown desert soils with *kankar*. VIII. Flat river valley: 19 — with xerophitic reverine forest on loam to medium heavy alluvial soils; 20 — with xeromorphic reverine thorn forests on sandy alluvial soils.

---

plant formations increases toward the west. It is characteristic that modern vegetation as a whole is more xerophitic than those which must grow under the existing natural conditions even in more dry districts. The fact that many dominant species of flora *(Acacia jasquemonti, Zizyphus nummularia, Euphorbia neriifolia)* belong to genuses which are usually spreading in monsoon forests and savannas of South India is the evidence of it. Really desertic are only *Capparis+* spp.[2] Xerophitisation of vegetation is connected not only with natural characteristics of landscape conditions but also with human activity. The everywhere thinning of mozophitic natural vegetation cover during the centres resulted replacement of climax forests and savannas by more unpretentiousness xerophitic species.

Desertification is a result of complicate interaction of natural and anthropogenic prerequisites. Study area is used by man about 7,000 years. Year by year pressure on land increases following population growth. For estimation of human activity it is necessary to take into consideration the specific types of economic landuse. Satellite imagery gives the possibility to determine it's main types (Fig. 24.2). Leading types of landuse are arable lands and pastures. Forests (Aravalli range and residual mountains) and water objects have seating spreading. Rainfed and irrigated croplands are characteristic for arable lands class. They clearly stand exposed on SI by colour. Dry-farmed lands occupy vast areas joining into great massives, and irrigated lands scattered by local plots along river beds, near storage pools, numerous ponds and "tanks". Intensive ploughing of marginal lands often of little use leads to their rapid degradation. These lands subjected to the processes of growth of sand fraction in humus horizon of soil, loss of fertility, reducing to fragments of soil particles and other. They become the plots with rapid development of desertification. Drawing of arable contours is small-spotted and it is accounted by the character of landownership – about 52 per cent of all arable plots have an area less than two hectares. Irrigation itself is not the cause of desertification, but indirectly can lead to secondary salinisation of soils, changing of their properties, waterlogging etc.

Arable lands occupy practically all western piedmont plains and space of socle plain in the east. Arable massives to a some extent are typical of all landscapes. The smallest arable areas are on more dry accumulative-deflation old alluvial plains. Connection of irrigated lands with certain landscape is expressed weaker due to its tie to water sources.

Pastures on SI are clearly distinguished by light plots of different intensity depending on their trapleness. Pastoralism is most typical for landscapes of alluvial plains. Often the growth of arable areas goes on owing to reducing of grazinglands under conditions of increasing livestock population, that leads to sharp growth of pressure on pasture areas and to decrease of pasture productivity. The plots of overgrazing are distinguished on SI by whitish tone. Routes of animal motion and trample down areas around wells are visible especially clearly. Negative consequences of overgrazing leading to desertification development are well-known.[10,20]

Primary natural vegetation in this region practically everywhere is reduced as a result of felling for building, fuel and fodder purposes. Many species are used by man for food especially in the years of droughts.[11] Plots of entire glades and forest pastures with fellings of middle intensity are visible on SI within Aravalli range on the general intensive red background which indicate forest vegetation. Such zones are found by general lightness of the tone. For landscapes of peneplain and the greatest part of piedmont plains complicated combinations of different types of landuse are typical.

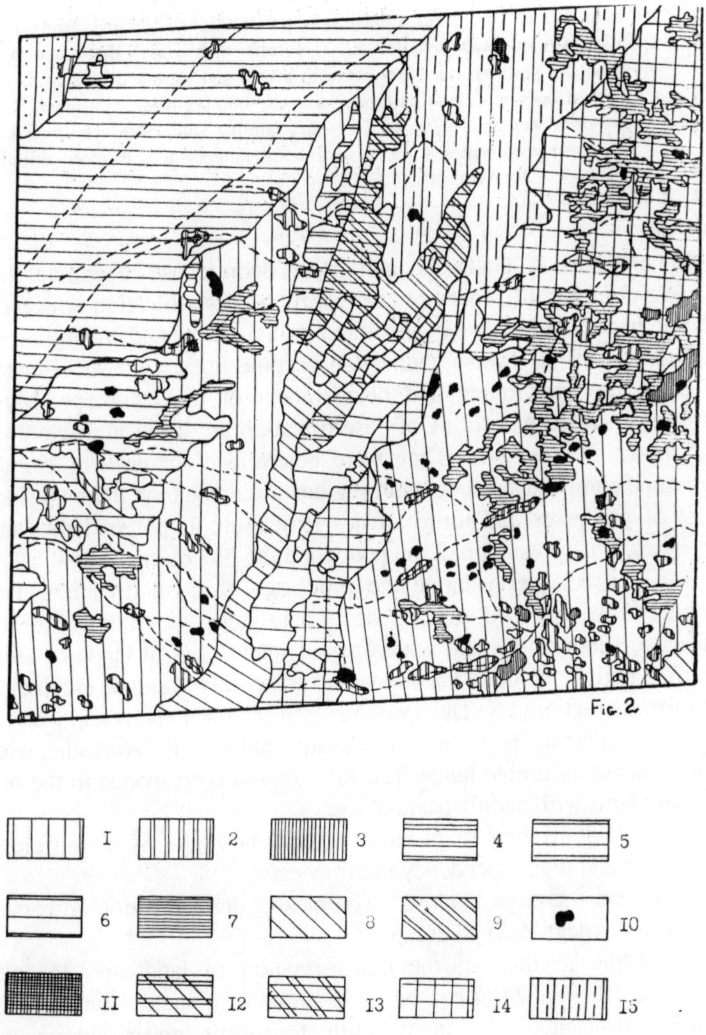

FIG. 24.2: Map of landuse of central Rajasthan (central part of Aravalli range). I. Arable lands: 1 — dry-farming croplands, 2 — irrigated croplands, 3 — lands under paddy. II. Pastures: 4 — grass, shrub and open woodland pastures, 5 — dry steppe and semi-desert rangelands, 6 — pastures with arable lands, 7 —

pastures along gullies and gorges. III. Forest management: 8 — forests and woodlands, 9 — forests with arable and slash — and burn land. IV Open water objects: 10 — lakes, storage pols, ponds, "tanks". V. Urban lands: 11 — cities. Complicated landuse: 12 — forests with grass, shrub and open woodland pastures, 13 — forests with arable and slash-and-burn lands and plots of open woodland pastures, 14 — dry-farming lands with grass, shrub and open woodland pastures, 15 — dry-farming lands with open woodland and dry steppe pastures, 16 — semi-desert rangelands with plots of arable lands, 17 — dry river beds.

Peculiarities of landscape and it's economic use determine that specific complex of processes and forms of desertification which on the one hand is derivative of them and on the other hand submitting to the mechanism of reverse connection become itself a powerful factor affecting the functioning and development of the landscape. Main leading processes and forms of desertification are listed in Table 24.1 and shown in Fig. 24.3. They were determined for each of twenty landscape complexes. Three main regions are notable on the map (Fig. 24.3) of processes and forms of desertification: 1. preponderate and processes of deforestation, 2. preponderate the processes of rapid water erosion and 3.) preponderate the processes of rapid deflation. The boundaries of exposed regions coordinate both with boundaries of landscape subdivisions and with areas of main types of landuse. Thus the first region includes the Aravalli system and residual mountains with forests and grades. The second region includes plains and plateau of the platform part and piedmont plains of Aravalli with predominance of arable lands. The third region corresponds to the old alluvial plains with mainly pastural use.

Analysis of all three maps based on deciphering of SI reveals close correlation and interdependency in the system:

landscape – type of landuse – complex of desertification processes – "deserted" (degraded) landscape.

According to this scheme transformation of landscape 16 into landscape 18 (Fig. 24.1) can be an example of a such chain. So this changed landscape can in its turn transform under subsequent development and intensification of human activity into a new more degraded state, i.e. become the beginning element of next chain.

For purposes of quantitative estimation of desertification some parameters of this process were determined by analysis of SI and literatures materials. Internal danger and present state of desertification were determined by scale worked out in the Institute of Deserts,

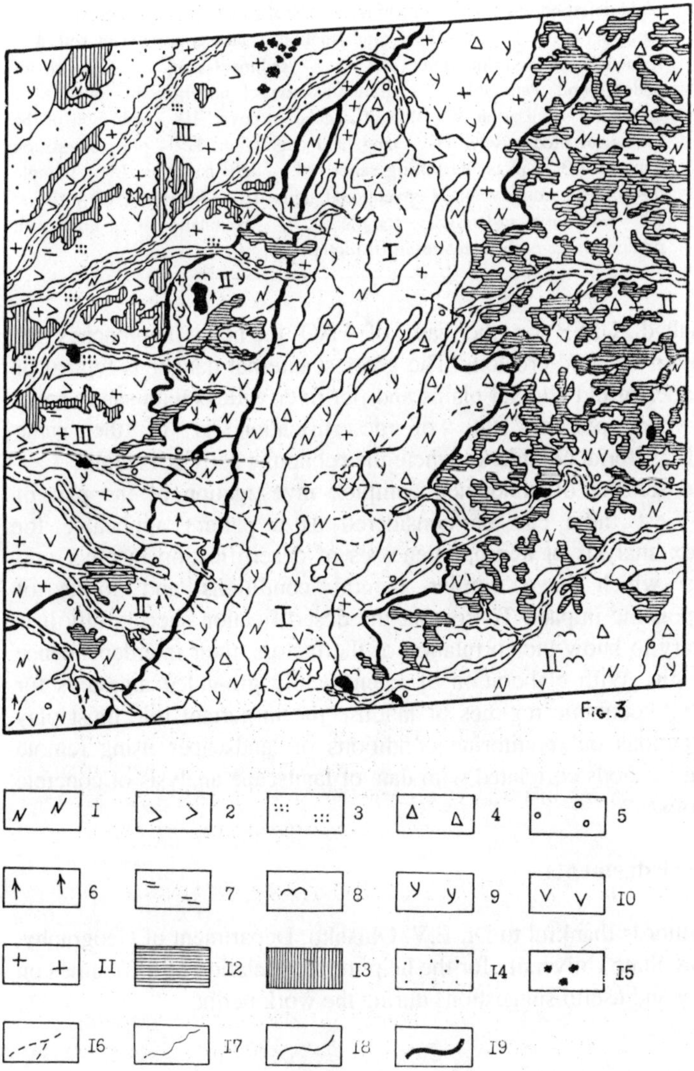

FIG. 24.3: Processes and forms of desertification in the landscapes of central Rajasthan (central part of Aravalli range). Areas with predominance of different desertification processes: I – preponderate the process of deforestation; II – preponderate the process of rapid erosion; III – preponderate the process of

rapid deflation. Processes and forms of desertification: 1 — rapid erosion, 2 — rapid deflation, 3 — growth of sand fraction in humus horizon of soil, 4 — reduction of concentration flow, 5 — sedimentation, 6 — uprise of underground watertable, 7 — mineralisation of underground water, 8 — secondary salinisation, 9 — thin of vegetation cover, 10 — reduction of the biological productivity, 11 — rangeland degression, 12 — plots of erosion network, 13 — plots of shifting sands, 14 — deflated areas, 15 — trample down areas around wells. Conventional signs: 16 — dry rivers, 17 — rivers; 18 — boundaries of landscape complexes, 19 — boundaries of the areas with predominance of different desertification processes.

Ashkhabad, USSR on the base of FAO methodic approaches for estimation of wind erosion.[7] The greatest internal danger is typical of landscapes of old alluvial plains and flat river valleys with sandy soils. The worst present state of desertification is in the same natural-territorial complexes where the economic activity is high.

The scheme of conducted complex investigation of the part of North-West India can be considered as a pattern applicable for complex analysis of appropriatenesses of desertification in other arid regions which have similar natural conditions and forms of anthropogenic impact. To combat the desertification successfully it is necessary to know the correlation of the factors, their interdependence and origin. With elaboration of scientifically grounded measures for changing economic regimes of landuse the important role must play investigations on monitoring conditions of landscapes using remote sensing methods correlated with data of landscape analysis of concrete territories.

Acknowledgements

The author is thankful to Dr. E.V. Glushko, Department of Geography, Moscow State University for the help in interpretation of difficulties on imagery and useful suggestions during the work period.

NOTES

1. *Atlas of forest resources of India*. (1976): National Atlas Organisation, Calcutta.
2. Balgoveschensky E.M. (1977): The Thar Desert in Western Rajasthan. Izvestia Vsesousnogo Geographicheskogo Obschestva, Leningrad, vol.101, no.4.
3. Carte Internationale du tapi vegetal. (1977): Notice de la Feuille. Rajasthan. De l'Institute Francais de Pondicherry, Pondicherry, no.12.
4. Desert and arid lands, (1984): Ed. El-Baz f., Martinus Nijhoff Publ.

5. Iliinsky N.D., Obiralov A.I. (1986): Fostikov A.A. Photogrammetty and deciphereing of photographs. Nedra, Moscow.
6. Ivakhov B.M. (1986): Satellite imagery used to study soil cover. Problems of desert development, Ashkhabad, no.1.
7. Kharin N.G., Babaev A.G. (1982): Studying and forecast of desertification processes by materials of cosmic survey. In: *The forecast of anthropoecological situation with the help of space means*. Nauka, Leningrad.
8. Kharin N.G. (1984): Application of distant methods for desertification processes study in arid zone. UNO, Moscow.
9. Kontoboitseva I.S., Kravtsova V.I. (1984): Studying and mapping the anthropogenic effects on the nature in different areas of the USSR from cosmic photographs. Vestnik Moscovskogo Universiteta, Serie 5, Geography, no.6.
10. Mabbat J.A. (1985): Desertification of the world's rangelands. *Desertification Control Bulletin*, UNEP, no.12.
11. Malhotra S.P. Socio-economic structure of population in arid Rajasthan. CAZRI, Jodhpur.
12. Mann H.S. (1986): Arid land development in South Asia. In: *Arid land development and the combat against desertification: An integrated approach.* USSR Commission for UNEP, Moscow.
13. Map of Quarternary deposits of asiatic part of ESCAP region. 1:5,000,000, Moscow, 1982.
14. Mission to Earth: Landsat views the world. NASA, 1976.
15. Nikolaev V.A. (1979): Problems of regional landscape science. Moscow State University, Moscow.
16. Problems and Prospects of desertification control in the ESCAP region. UNEP, Bangkok, 1983.
17. Saxena S.K. (1977): Vegetation and its succession in the Indian desert. In *Desertification and its control.* ICAR, New Delhi.
18. Singh G. (1980): A geography of India. Progress, Moscow.
19. Soil map of the world, 1:5,000,000. FAO/UNESCO, Paris, vol.7, 1977.
20. Zonn I.S., Orlovsky N.S. (1986): Anthropogenic factors of desertification. In *Arid land development and the combat against desertification: An integrated approach.* USSR Commission for UNEP, Moscow.

TABLE 24.1: The Factors, Processes and Forms of Desertification in the Landscapes of Central Rajasthan

Land-scape NN	Types of Landuse	Desertification		
		Factors	Processes	Forms
1, 3, 4.	Woods and forests	deficiency of precipitation; seasonal unevenness of precipitation; shallow underdeveloped soils; steepness slopes.	rapid erosion; thin of vegetation cover; reduction of the concentrated flow; cutting.	outcrops; soil and vegetation cover; degradation (replacement forests by shrub and grasses.
2, 5.	Forests with seats of arable and slash-and-burn cultivation; plots of open woodland pastures; forests.	deficiency of precipitation; seasonal unevenness of precipitation; elimination of vegetation; pasturage; close position of rocks.	thin of vegetation cover; rapid erosion; reduction of the concentrated flow; rangeland degression.	outcrops; soil and vegetation degradation; trample areas.
6, 7.	nonirrigated lands in combination with open woodland and semidesert pasture.	deficiency of precipitation; seasonal unevenness of precipitation; elimination of vegetation; pasturage; growth of arable areas; irrational use of water resources.	rapid erosion; reduction of the concentrated flow; rangeland degression; reduction of biological productivity	soil and vegetation degradation; trample areas; gullies.
8.	pastures with arable land; non-irrigated lands and shrub pastures.	deficiency of precipitation; seasonal unevenness of precipitation; elimination of vegetation; overgrazing.	rapid erosion; reduction of the concentrated flow; sedimentation; reduction of biological productivity; rangeland degression; local deflation.	erosion network; soil and vegetation degradation.
9.	open woodland, shrub and grass rangelands; nonirrigated lands.	deficiency of precipitation; seasonal unevenness of of precipitation; stepness of arable lands; light soils; deplition of water resources; growth of arable areas; overgrazing.	rapid erosion; rangeland degression; deplition of soils.	degraded soils; spotted and mosaic vegetation cover.

Contd

TABLE 24.1 — *Contd.*

Land-scape NN	Types of Landuse	Desertification		
		Factors	Processes	Forms
10.	nonirrigated land with open woodland and semidesert rangeland; arable and irrigated lands.	deficiency of precipitation seasonal unevenness of precipitation; deplition of water resources; elimination of vegetation; overgrazing.	rapid erosion; thin of vegetation cover; rangeland degression; reduction of soil productivity; sedimentation; reduction of concentrated flow salinisation.	erosion network; spotted and mosaic vegetation cover; salty plots.
11.	nonirrigated and irrigated lands.	deficiency of precipitation seasonal unevenness of precipitation; light soils; deplition of water resources.	rapid erosion; sedimentation; thin of vegetation cover; reduction of biological productivity; uprise of underground watertable; water mineralisation.	erosion network; spotted and mosaic vegetation cover.
12, 13.	nonirrigated and irrigated lands; lands under rise; pastures along gullies and gorges.	seasonal unevenness of presipitation; elimination of vegetation; deplition of water resources; close position of cristallic socle; heavy soil.	rapid erosion; reduction of biological productivity; secondary salanisation; reduction of concentrated flow; sedimentation.	widespread erosion network; salty areas; badlands; sporadic vegetation.
14.	nonirrigated lands with open woodland shrub and grass pastures.	deficiency of precipitation seasonal unevenness of precipitation; deplition of water resources; overgrazing.	rapid erosion; reduction of concentrated flow; rangeland degression; reduction of biological productivity.	widespread erosion network; trample areas; spotted vegetation cover.

Contd

TABLE 24.1 — *Contd.*

Landscape NN	Types of Landuse	Desertification		
		Factors	Processes	Forms
15.	nonirrigated and irrigated lands; rangelands with arable lands.	deficiency of precipitation; seasonal unevenness of presipitation; droughts; ploughing of marginal land sandy soils; overgrazing; mineralised underground water.	rapid erosion; rapid reflection; uprise of underground water level; thin of vegetation cover; rangeland degression; blowing of upper soil layer.	sparse vegetation cover; deflated areas; trample areas.
16, 17, 18.	dry steppe and semidesert rangelands.	deficiency of precipitation; seasonal unevenness of precipitation; elimination of vegetation; overgrasing sand and dust storms; droughts; deplition of water resources; *kankar* horizon in soil; sandy soils; salt underground water.	rapid deflation; rapid erosion; reduction of the concentrated flow; sedimentation; growth of sandy fraction in soil humus horizon; underground water mineralisation.	shifting sands; deflated areas; trample areas; degraded soils and vegetation cover; sparse spotted vegetation.
19.	irrigated lands	elimination of vegetation; close position of crystallic socle; heavy soils; irrational use of water resources.	reduction of the concentrated flow; rapid erosion secondary salanisation; sedimentation.	erosion network; trample areas; sparse vegetation; areas of salin soils.
20.	dificated lands; semidesert pastures.	deficiency of precipitation; seasonal unevenness of precipitation; sporadic flow; elimination of vegetation; overgrazing; irrational use of water resources.	rapid erosion; rapid deflation; underground water mineralisation; rangeland degression; sedimentation; blowing of sandy soils.	shifting sands; trample areas; gullies; erosion network; sparse spotted vegetation.

25

Application of Remote Sensing Data for Geomorphological and Environmental Conservation Studies

M.S. NATHAWAT, M.P. PUNIA, R.R. SANI and SANJAY RAJ

Introduction

THE ABNORMAL high rate of growth during the present century and lack of management during the British period have resulted in enormous damage to our landscape. The problem has been highly accentuated in areas where a delicate balance existed between the environment and land potential. In order to restore the productivity of the land and amelioration in the ecological status of the area, intensive data have to be collected within short period of time. Remote sensing technique could provide such opportunity to those who are planning for environmental conservation and not only improving the overall productivity but also preservation of environmental quality.

The present study has demonstrated that the geomorphological information about an area is an essential base for the integrated survey and mapping. The study of geomorphology deals mainly with the processes acted on the earth surface, responsible for the land development. This study is vital for the preparation of soil maps, hydromorphological maps and to study the ecological status in the area. While man is capable of inducing significant alteration in the environment, environment is endowed with multidimensional system of complex interrelationships in a continuing state of change. The studies, therefore, should be equally capable of covering wide range of variables both temporal and spatial. Modern remote sensing techniques do have such capability.

Objectives

Objectives of this particular study aim to:

1. to assess the comparative geomorphological contents of MSS and TM data.
2. to delineate the land-use/land cover on various geomorphological features.
3. to study the trends of land transformation done by natural and human agents.

Location and Extent

At 1:1 M scale between latitudes 23° 35′ and 25° 22′ N and longitudes 72° 35′ and 74° 22′ E the area covered is about 34000 sq km. All mapping has been at 1: 2,50,000 scale. It covers parts of Udaipur, Sirohi, Chittorgarh and Banswara districts of Rajasthan and Mehsana district of Gujarat.

Methodology

The landsat satellite scene of MSS and TM Row 148 and path 043 at 1:1 m scale was enlarged at 1:2,50,000 scale using Procom-2 and density slicing was done using Image Analyser on band number 7 of TM. The enlarged scene was interpreted visually for geomorphology, surface water and environmental conservation study. The images were analysed and pre-field maps were prepared for the above themes. Images were scanned using illuminated magnifier (Dianascane). The features were identified and delineated on an overlay. Subsequently based on limited ground check and background knowledge of area, reinterpretation was done, which included few revision of boundaries. The various resource data integrated include geology, geomorphology, hydrogeology, soils and land use for generating environmental conservation maps.

Results and Discussions

Geomorphologically the study area represents a fairly rugged relief with structural hill ranges, linear and curvilinear ridges of quartzites and plateau with intermontane valleys in the eastern sector. These are bordered by piedmont and pedimented areas, as well as pediplains to the east and west of the hill relief. A number of water bodies occupy local depressions. The western part of the area depicts a vast pediplain, the monotony of which is broken by the hill relief around Sirohi, Mount Abu and a few dotted hills north of Erinpura. The structural hill ranges trend more or less NNE-SSW. The main drainage system in the area flows southernly or north-westerly and comprises of the Jawai and Sabarmati river systems and its tributaries. Elevation ranges from around 308 m to 1097 m above mean sea level in the area.

The following six predominant geomorphic units have been mapped in the area.

1. Denudational hill,
2. Structural hill,
3. Residual hill,
4. Pediment,
5. Pediplain and
6. Flood plain.

Denudational Hill

The denudational hills are wide spread in the southern part and taper towards the northern part, having the maximum relief in this area. These are marked by sharp crest lines with narrow intervening valleys where the down cutting of the rivers are more predominant. The wider valleys are marked by valley fill deposits.

Structural Hill

The structural hills show clear dip slope and steep escarpment slopes, occurring in the western part of the study area. The land cover types is of scrub and open forest.

Residual Hill

The residual hills are of still lower altitude when compared to other hills and occur mostly as isolated outcrops or inselbergs, though their continuity in a linear or curvilinear fashion gives indication that they are structurally controlled. These are mostly barren or with sparse vegetation.

Pediment

In the pediment unit, the areas of sheet erosion are marked by a high reflectance where the soil is getting eroded due to sheet flow of rain water and can be classified as wate land.

The buried pediment comprises of thick weathered rock surfaces with good potential groundwater. It can be further classified into two categories as deep buried pediment and shallow buried pediment.

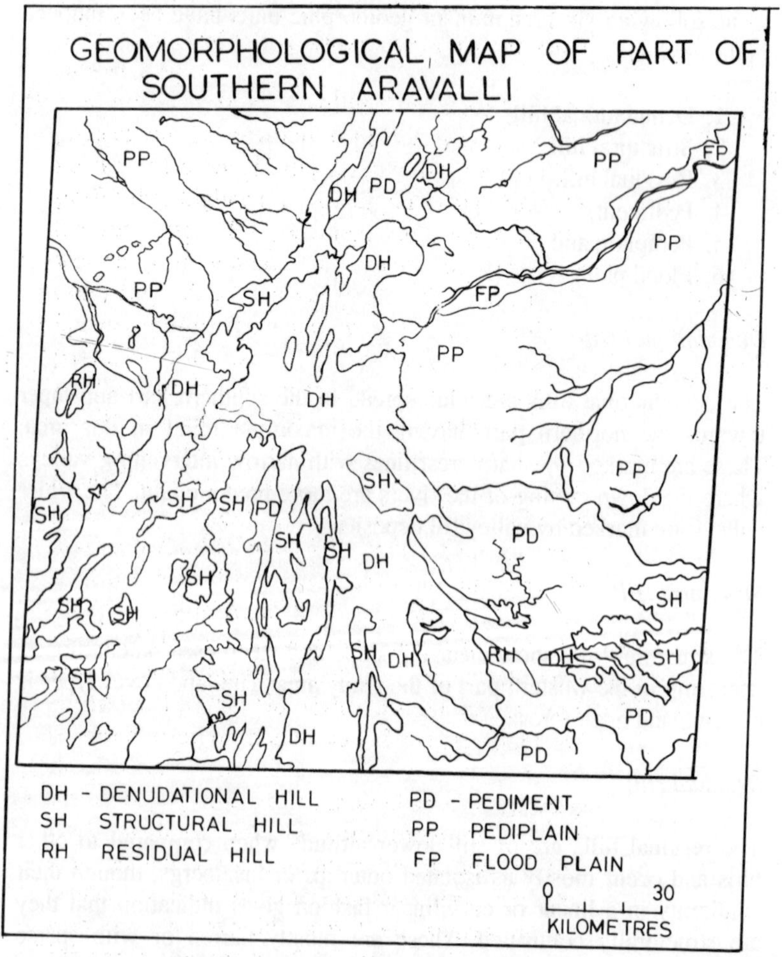

FIG. 25.1: Geomorphological Map: Part of Southern Aravalli.

Pediplain

The pediplain consisting of several pediments, a rock cut surface at the bottom of the relief forming generally an undulating topography. They are marked by the typical yellowish tone in the FCC.

Flood Plain

The flood plain is restricted to the seasonal ephimeral rivers. The upper reaches of rivers are associated with tanks/bands wherever the topography is beneficial to form a small reservoir.

Since landforms form the basic units which will have relationship with soils, landuse and groundwater, they form and ideal base for preparing an integrated resource map. The concept of land system will, therefore, show the interdisciplinary relationship, and it is recommended that the land system units as observed in the area may be used while preparing an integrated environmental conservation map.

While interpretation of satellite images of different scales provided information on regional distribution of landforms, the ground truth information collected during the field checks has added further information particularly on the environmental aspects. The skyline of the relief areas indicates that the relief of the area underwent prolonged denudation resulting in planation surfaces. This is evident from almost horizontal skyline that can be observed in some parts of the area. Further it is also evident that these planation surfaces have further undergone dissection, since in many places the relief is made up to the subrounded to rounded crest line and smooth, concave hill slopes.

A vertical profile of the relief area indicates the close relationship between the structure and topography since slopes are generally related to dip slopes giving rise to hogback and questa topography. However differential erosion resulting due to variation in hardness of the rocks is also evident in hard rocks standing but prominently with almost no soil cover, whereas the soft rocks are usually covered by soil and surface wash material. From the nature of the soil profiles seen in the pediplain, it is further evident that sheet wash process has been the predominant process in the past indicating that the area had lacked vegetation in the past.

The stony soils as well as red profiles of the soils bear testimony to the fact that the predominantly dry conditions existed in the past. However, at the same time, the occurrence of the narrow piedmont zone in front of relief areas with first order drainage basins in the relief are no longer actively dissecting. This is indicative of a wet period in the past before the present day dry conditions. Even in the present day poor vegetation cover in the most of the relief areas with practically no under growth and formation of terracettes on the slopes indicate a definite evidence of overgrazing and sheet erosion, removing the top soil faster than the rate of the soil development.

Profiles of regolith indicate the shallow buried pediment in many areas. On the whole environmental aspects indicate that the area is undergoing active sheet and rill erosions exposing bedrock areas and depleting the areas covered by soils. This trend, if left unchecked, will further degrade the environmental conditions in the area.

The study indicates a poor forest cover in the area. Wherever forests are existing they could only be identified as very poor with scrubs. Though there is a very large area under rainfed agriculture, there are mostly low hills and piedmont and pediment. A large chunck of land could be taken out of this for plantation purposes with the objective to conserve diversity and integrity of landscape.

What is required urgently is to actively grow appropriate vegetation on areas having sufficient soil cover so that the vegetation in turn will protect the exposed rocky areas from further degradation.

Conclusions

Satellite data employed in this study provided better tonal and textural variations and sharpness. This facilitated delineation of geomorphic features with greater ease. These satellite data offer opportunities to arrive at accurate land evaluation and assessment of environmental reliability. The use of satellite data for preparing such maps is quicker than conventional methods.

REFERENCES

Culingfard, R.A., Davidson, D.A., & Lewin, J. (1980): *Timescales in Geomorphology*, John Wiley & Sons, New York.

Douglas, I. & Spencer, T. (1985): *Environmental Change and Tropical Geomorphology*, George Allen & Unwin, London.

Rhodes W. Fairbridge (1968): *The Encyclopaedia of Geomorphology*, Dowden, Hutchinson & Ross, Inc., Pennsylvania.

Robert N. Calwell (1983): *Manual of Remote Sensing*. American Society of Photogrammetry, Falls Church, Virginia.

Sen, D., (1972): Geomorphology of the Aravalli range, Rajasthan and reinterpretation of residual erosion surfaces, *Geographical Review of India*, Vol. XXXIV, No. 3, pp. 265-286.

Verstappen, H. Th. (1983): *Applied Geomorphology*, Elsevier Science Publishers, Amsterdam

26

Remote Sensing as a Tool for Assessing Environmental Effects, Present State, Changes and Ecological Significance of Reservoir System

MAHENDRA SINGH NATHAWAT, ANIRUDH SAHNI, DHARMENDRA SHARMA and A.K. SINHA

Introduction

DIFFERENT LAKES and coastal sea areas are amongst the few most essential environments forming a natural surrounding for almost all activities of men. Hence, water management has become one of the most important elements for our life now.

This means that aquatic areas are the object of various and often conflicting demands of utilisation. In this field, it is necessary to know the present status of water qualities periodically in order to manage the water resources. Research, investigation and planning of water areas is, therefore, essential for recreational and residential use and as a source of livelihood and raw water or as a dumping place for waste water, all require accurate information, some of which can in principle be obtained and prepared by satellite remote sensing which is a powerful tool in this field.

Remote sensing data are increasingly used in various assessment, management of natural resources and environment such as climate, oceanography, marine resources, water resources, landuse analysis, agriculture and soil information, forest and ecology, geological applications, human settlement analysis and engineering applications etc.

The surface water present in rivers, lakes and swamps is, in principle, readily observable on imageries but the important seasonal and secular variations of its availability necessitate careful study of the characteristics of form, gradient and roughness of the riverbed. This variability and the fact that water may be required at localities other than those where it is abundant, may necessitate engineering structures of various kinds so as to make optimum use of the surface water resources.

Geomorphological factors also play a role in photo geological, or pedological surveys, as an aid in establishing lithology, structure, aquifers and many other matters of hydrological relevance. Data on surface water features can also be easily and accurately observed and measured on these kinds of imagery. It is evident from the aforesaid that geomorphological studies for hydrological purposes should stress the role of landforms and other geomorphological factors in the broadest sense and should ultimately lead to an appreciation of the environmental situation in which the water occurs.

Presently various remote sensing application in water resources are:–

1. Synoptic meteorology analysis.
2. Snow cover mapping and snow melt runoff forecasting.
3. Rainfall estimation.
4. Surface water mapping and inventory.
5. Flood inundation mapping and damage assessment.
6. Reservoir surface water spread monitoring.
7. Study of river behavioural changes.
8. Water quality monitoring and reservoir sedimentation analysis.
9. Watershed characterisation.
10. Bathymetry and hydrographic survey.
11. Landuse/cover in the watershed.
12. Estuary and wetland surveys.
13. Groundwater potential zone mapping.
14. Canal command area mapping.
15. Identification of erosion prone areas.
16. Estimation of evapotranspiration.
17. Soil moisture determination.

In brief, remotely-sensed data offer valuable information for better management of water resources and hydrologic studies. The present study describes some attempts to apply aquatic remote sensing to the preparation of parametric map such as presentation, quantitative evaluations and time-related investigation of Ramgarh lake, Jaipur. It is the largest water source for Jaipur city which has an estimated population of more than 20 lakhs. The regulation of this lake is a continuous task of undiminishing importance.

Objectives

The aims of the study was to know the present state and the late changes of ecozones and land cover of the watershed and the water areas of Ramgarh lake which has a large catchment of 760 sq km, situated 30 km north-east of Jaipur among the hills in the gorge through which the Banganga river passes before emerging on the plains. There are several schemes both to bring out and protect this natural media where changes during the past fifteen years became significant owing to supply of drinking water, fresh water fishing, cultivation and development of water sports and recreation. The remote sensing seems a good tool to give cartographical, quantitative and qualitative data as an aid to the decision-makers. Keeping this in view, objective of the study in therefore, framed into the following steps:

1. To extract physical information of landuse-landcover and stream network of the watershed from different satellites data sources: Landsat – TM, MSS and TRS-1A, LISS II, FCC image and single band paper products.
2. To map and highlight the degradational part of the basin.
3. To ascertain the efficiency of surface water mapping using Landsat TM, MSS and IRS-1A, LISS II imageries.

Methodology

The satellite data include information of aquatic areas down to the secchi disc depth. The green and yellow wavelengths of channel 1 indicate variations with a few uppermost metres in water features, even down to tens of metres within certain open ocean areas. The red radiation of channel 2 usually comes from within the uppermost one metre layer, and the near infra-red channels 3 and 4 include information for within the uppermost centimetres or millimetres respectively. In addition, the uppermost part of the water column is best represented within all recorded channels.

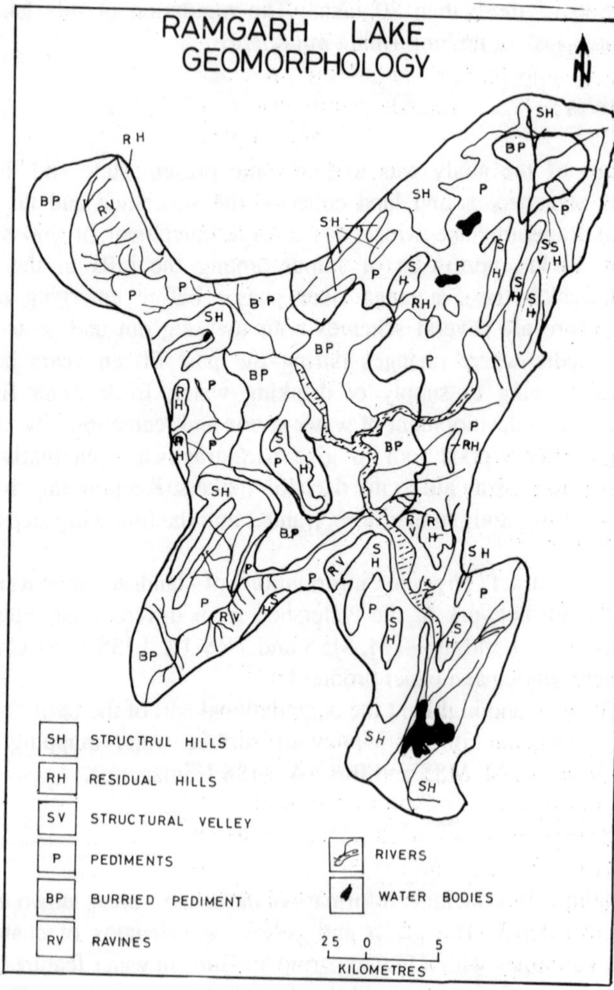

FIG. 26.1: Ramgarh Lake: Geomorphology.

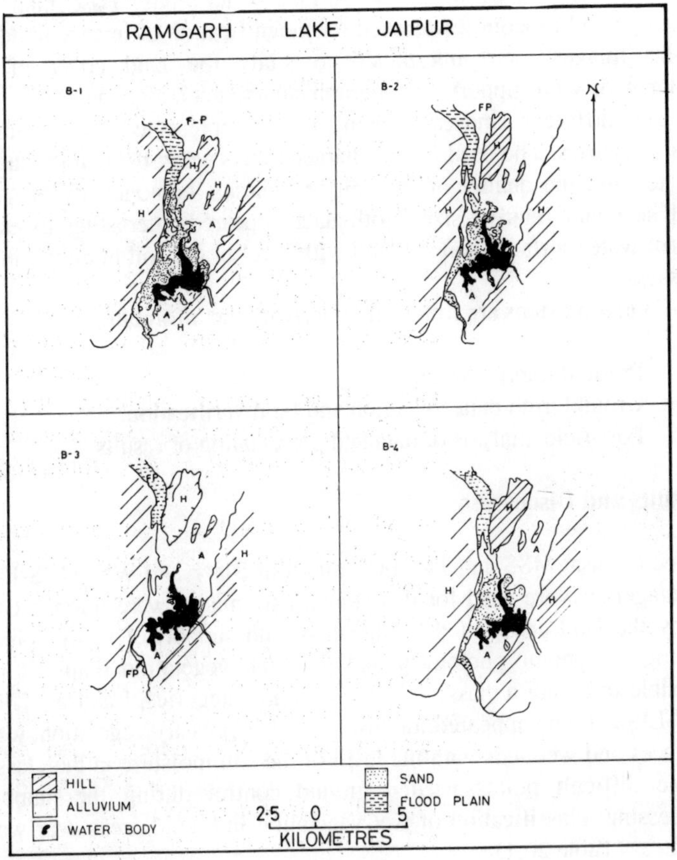

FIG. 26.2: Comparison of Part of IRS LISS-II Image bands 1 to 4 showing Ramgarh lake. Band 1, panetrates the water to show the greatest amount of turbidity; Band 2 high turbidity with more distinct shoreline; Band 3 shoreline clearly delineated, no water penetration; Turbidity pattron less visible but sediment visible, slight water penetration

The main task of the present work is the interpretation of surface water body and the landforms, landuse/landcover in the area of interest.

Landsat TM, MSS and IRS-1A, LISS II and false colour prints on 1:1,000,000 scales were used in conjunction with the Survey of India's corresponding topographical maps on scales 1: 50000 and 1: 250,000 to recognise and delineate the surface water body. Base map was prepared and procom-2 was used for magnification of images up to the scale. Images were interpreted to study the land cover of the watersheds with support of topographical and geological maps.

The different landscape elements are represented by the image characteristics. The used image characteristics were: the colour classes, the texture, the quality of the unit boundaries, homogeneity of shape and size and relative size. Both data appeared interesting to easily detect water surfaces and wetland with level of the soil humidity in the classes.

Following steps are followed in deriving the information:

- Prefield interpretation.
- Ground truth data collection and field verification.
- Post-field analysis of data and presentation of results.

Results and Discussion

Three Landsat MSS and TM, FCC and four Black and White IRS-LISS II imageries were used for this study. The images were interpreted to study the land cover of the watershed with support of topographical, geological, geomorphological and potential vegetation maps. It was possible to obtain a classified map with all categories. Landsat, TM & IRS-LISS II data appeared interesting to detect easily vegetation, water surfaces and wetlands with the level of the soil moisture in the classes. Some difficult points needed ground control during the mapping processing: classification of wet vegetation in humid areas near water surfaces (Table 26.1)

TABLE 26.1: Qualitative Estimate of Relative Turbidity of Water from IRS-1A Image

Relative turbidity	Tone of image				Hue of colour composite
	Band 1	Band 2	Band 3	Band 4	
None	Dark	Dark	Black	Black	Black
Slight	Medium	Dark	Black	Black	Dark Blue
Moderate	Light	Medium	Black	Dark	Medium Blue
Heavy	Light	Light	Dark	Medium	Light Blue
Very Heavy	Light	Light	Medium	Light	White

On analysing the data we find that in the past few years there has been a rapid decrease in the level of water in the lake mainly due to shortage of rainfall in the catchment areas and deforestation leading into formation of ravines and consequently transportation of silt into the lake. It has been recorded that 5 metres of sand has been deposited till now. Today the situation is that hardly 6 metres of water is present which is insufficient to fulfil the domestic needs of the city people. Secondly though less important than the first is that Ramgarh was chosen a place for water sports, Asian games were held in 1982 and during that time it was proposed to develop it into a permanent sporting ground, but then looking to the unfavourable conditions developing the idea has been given up.

The changes since 1972-89 were distinguished by comparison of interpreted TM & IRS-LISS II data with corresponding MSS data. Between the two dates a decreasing precipitation and reduction in surface water resulted in the decrease of agricultural activity and water sports.

Conclusion

The study on Ramgarh wetland has shown the high interest of the Bands of Landsat & IRS data to know at 1:50,000 scale, the State of the large wetlands, land cover with details about the vegetations of humid media. It gives the possibility also to discriminate the characteristics of depth, aquatic vegetation and turbidity of water areas.

The observation of the land cover changes during the 1972-89 period in Ramgarh lake and its vicinity by satellite remote sensing was possible and did not present too many difficulties.

REFERENCES

Anirudh Sahni (1987): "Reservoir system in Rajasthan and its Ecological significance", An unpublished dissertation submitted to Indira Gandhi Centre for HEEPS, University of Rajasthan, Jaipur.

Reid, G.K. and Wood, R.D. (1976): "Ecology of Inland Water and Estuaries II", Edn. D. Van Nostrand Co., New York.

Rajasthan Pollution Prevention and Control Board (1986): "Environmental Status of Lakes in Rajasthan".

Ross, D.I. and V. Singhroy (1983): "The Application of Remote Sensing Technology to the Environmental Assessment Process", prepared for Workshop on New Directions in Environmental Assessment : The Canadian Experience, Toronto, Ontario.

Verstappen, H.Th. (1983): "Applied Geomorphology", Elsevier Science Publishers. Amsterdam.

27

A Conspicuous Low Stand of Sea Level in the Inner Shelf Zone off Visakhapatnam (East) Coast, Andhra Pradesh

K.N. PRUDHVI RAJU and AHMED RAZA KHAN

The Coast of Study

THE EAST coast in north coastal Andhra Pradesh along a length of about 120 km from Revu Polavaram to Mukkam in the districts of Visakhapatnam and Vizianagaram (Fig. 27.1) is rocky with a few promontories protruding into the sea; a few ridges run along the coast and a few hills lie abutting the waterline forming irregular elongated bays in between the rocky headlands, often containing beaches and other related forms of accumulation up to a few hundreds of metres backed by plains and/or hills. A pair of comparatively major rivers along with a couple of minor rivers and a few smaller coastal consequents and tidal creeks interrupt and create marshy, swampy and mangrove environments along this stretch of the coast.

The central portion of the eastern Ghats between the Godavari and Mahanadi rivers with a mountainous terrain rising to a maximum height of 1620 metres above m.s.l. and linear ridges, a few hills and hillocks in the altitudinal range of 30 to 600 metres within the plains towards the coast, constitute the hinterland of the coast of present study.

The principal rock types along as well as in the hinterland of the coast are the Khondalite (quartz-felspar-garnet-sillimanite genisses) and Charnockite (hypersthene bearing gneisses and granulites) Groups of Archaen rocks. Quartz viens and pegmatites run and cut along and across these rocks here and there. The rock formation trend in a

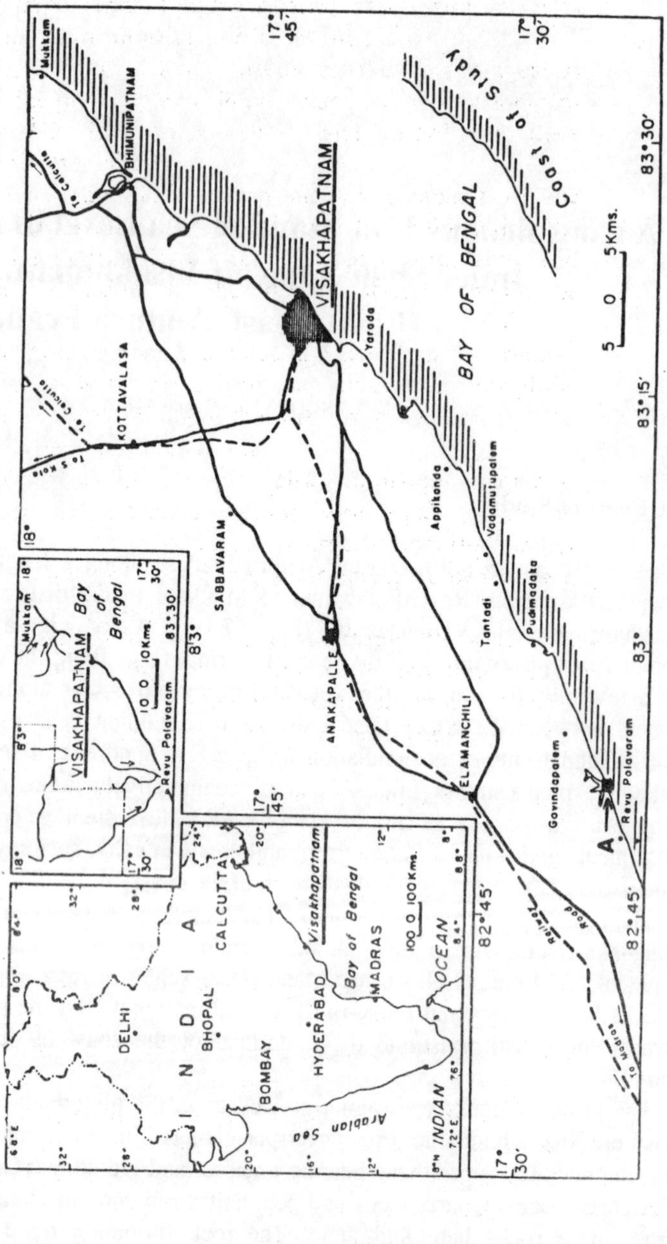
VISAKHAPATNAM
BHIMUNIPATNAM
Mukkam
KOTTAVALASA
To Calcutta
SABBAVARAM
ANAKAPALLE
ELAMANCHILI
Yarada
Appikonda
Vadamutupalam
Tantadi
Pudimadaka
Govindapalem
Revu Polavaram
BAY OF BENGAL
Coast of Study
5 0 5 Kms.
To Madras
Railway
Road
A
INDIA
DELHI
BHOPAL
CALCUTTA
BOMBAY
HYDERABAD
MADRAS
Visakhapatnam
Bay of Bengal
Arabian Sea
INDIAN OCEAN
100 0 100 Kms.
10 0 10 Kms.
18°
17° 45'
17° 30'
82° 45'
83°
83° 15'
83° 30'

general NE-SW direction dipping due southeast with deviations in strike towards NNE-SSW, N-S and E-W dipping due northwest, west and south. The folds are predominantly non-plunging to gently plunging tight (overturned) isoclinal folds with large scale basin and dome structures here and there (Natarajan and Nanda, 1981).

The Study

It is well-known that the sea level locally, regionally and globally was lower than the present level though at varying depths and times in the past. It is doubtful whether the inferred Ordovician and Permian transgressions (Fairbridge, 1960) and more or less confirmed Cretaceous transgression, started from a level lower than that of the present. The falling sea levels during the middle tertiary and then an upswing higher than today by late Tertiary are resolved beyond doubt (Tanner, 1968). The world-wide regression and low sea levels during Quarternary (from Pleistocene till the beginning of Holocene) are generally accepted. The sea level was lower than today several times during Quarternary more than the number of times worked out earlier (Shackleton and Opdyke, 1973; Moore, 1982 and Chappel and Shackleton, 1986). The Holocene fluctuations as worked out by Fairbridge (1961) are yet to be confirmed beyond doubt.

Studying indicating substantive evidences of low stands of sea level in the shallow region of the inner shelf zone along the east and west coasts of India are very few. As for the west coast, the Gaveshani bank of corals from Konkan at 16.75 metres depth are dated to be of early Holocene–between 8,000 and 10,000 years BP (Nair and Qasim, 1979). Patel *et al.* (1984) recognised late Upper Pleistocene regression up to 20 metres from the present level along Mahi-Tapi segment of Gujarat coast. The sea level curve drawn by Kale and Rajaguru (1985) for west coast shows four still stands between 50 and 100 metres from the present level.

Carbon 14 dating revealed Holocene low levels in the Cauvery delta at 30 metres and 25 metres below the present level in the present coastal part and at a location 50 km inland respectively (Sadakata, 1980). Banerjee and Sen (1987) through dating of subsurface sediments and peat material identified a low sea level at 2-7 metres below the present during 6,500-7,000 years ago in the West Bengal plains.

Off the coast of Visakhapatnam, Mahadevan and Rao (1954) located relict sediments laid down in an environment of lower seas at

40-54 metres depth from the present level. Low levels at 20 to 23 metres depth around 7,000-8,000 years BP in the Visakhapatnam harbour area are reported by Prudhvi Raju and Vaidyanathan (1978). Murthy *et al.* (1987) traced carbonate reefs between 50-60 metres showing linear trend parallel to the shelf edge and concluded that they might represent ancient shorelines.

An attempt has been made in the present study to find out the low stands of sea level in the near shore region of the inner shelf zone off Visakhapatnam (east) coast. The echo-sounding surveys along this stretch of the coast reflected very smooth and featureless bottom topography over the inner and mid-shelf zone (Rao *et al.*, 1980). The study is based on field investigation at the confluence of the Sarada river (A in Fig. 27.1). Also, Naval Hydrographic Chart No. 3002 served as another tool of the present study. The area of field investigation falls outside the Chart area. The Chart gives information of soundings of water depth in fathoms in two parts, one off Visakhapatnam proper in the immediate vicinity of the harbour entrance with soundings up to 18 fathoms (32.4 metres) and the other with soundings up to 35 fathoms (63 metres) covering a larger area north and south of Visakhapatnam proper.

Fathomlines at one fathom (1.8 metres) interval are interpolated from the soundings in the Chart (Figs. 27.2 and 27.3). Then a number of cross profiles at regular and close intervals are taken across the contours and fathomlines.

An examination of fathomlines in Fig. 27.2 shows a general grouping of fathomlines from 1 to 6 (1.80 to 10.80 metres) excepting at the north breakers near harbour entrance channel. The gap between fathomlines 6 and 11 (10.80 and 19.80 metres) widens before narrowing down between 11 and 15 fathoms. The corresponding superposed profiles (Fig. 27.4a) show breaks in several profiles at 1, 4, 6 and 11 fathoms. The break at 6 and 11 fathoms by far is most common in most of the profiles. The fathomlines along a longer length of the coast (Fig. 27.3) north and south of Visakhapatnam proper show a clear clustering from 1 to 6 fathoms and spread somewhat wide further down up to 22 fathoms. The 22 fathomline marks a clear break and the inner shelf forms a gently sloping surface further down. The corresponding superposed profiles (Fig. 27.4b) show general breaks across most of the profiles at 6 and 11 fathoms (10.80 and 19.80 metres). An almost uninterrupted, uniform and steep topographic (cliff) face up to a depth of 6 fathoms (10.80 metres) can be observed from the profile pattern.

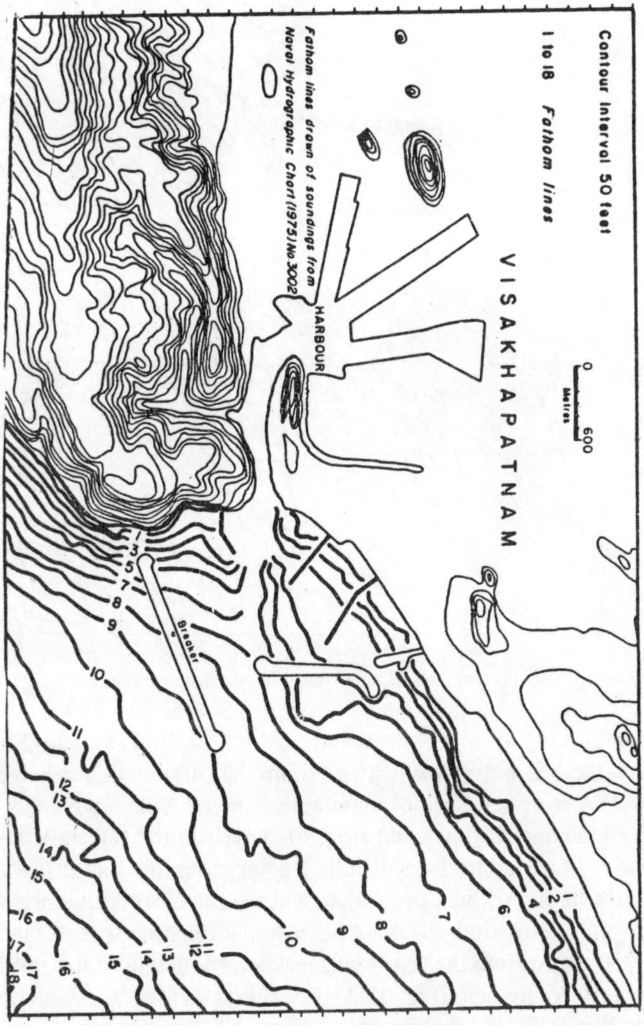

FIG. 27.2: Map showing fathomlines near Visakhapatnam Harbour Entrance .

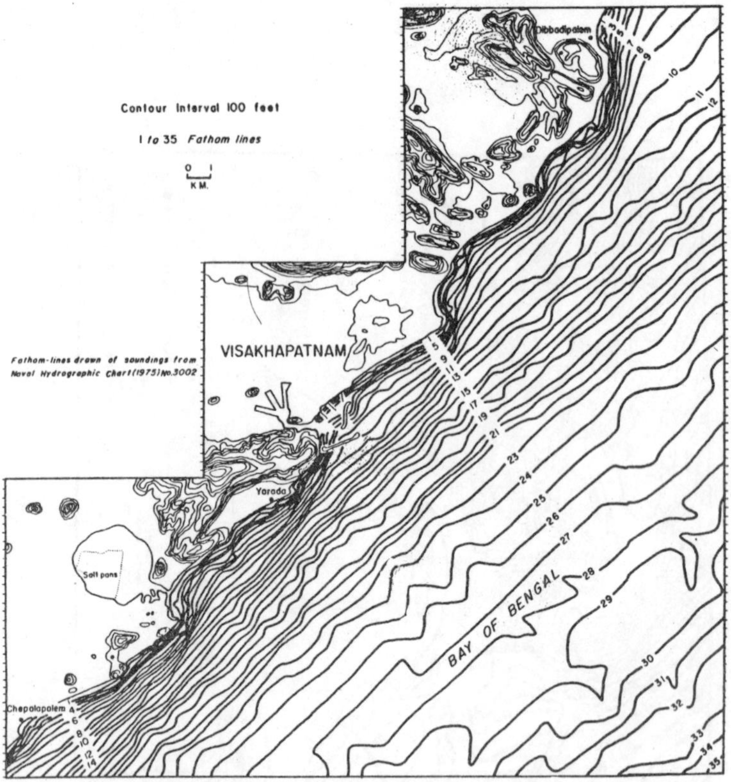

FIG. 27.3: Map showing Fathomlines off the Coast of Visakhapatnam Region.

The Sarada estuary at it's confluence (A in Fig. 27.1) is bound by two hills in the NE and SW which rise more than 100 metres in height above m.s.l. A wide sand spit of maximum 2 metres relief above m.s.l. formed by a number of low sand ridges is keyed to the NW hill. The spit extends almost up to the SW hill leaving a narrow gap of about 20-25 metres through which the Sarada and the sea meet each other. A drowned valley of a width of about 30 metres extending to a distance of about 250 metres into the sea from the seaward margin of the spit is a conspicuous feature here (Fig. 27.5). It's bottom is at about 10 metres depth from the MSL and the general surface on both sides of it, almost all along its length, which is rocky is at a depth of about 2 metres from m.s.l.

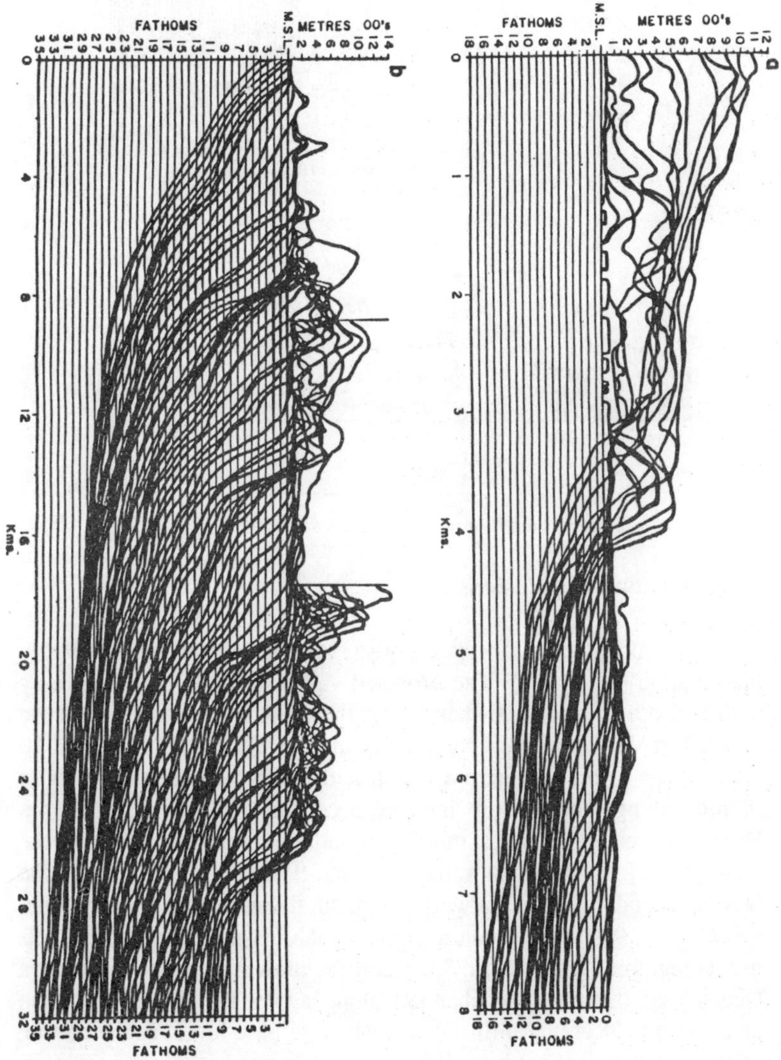

FIG. 27.4: Superposed Cross-profiles taken Across Contours and Fathomlines of Visakhapatnam Area (Proper) and Region.

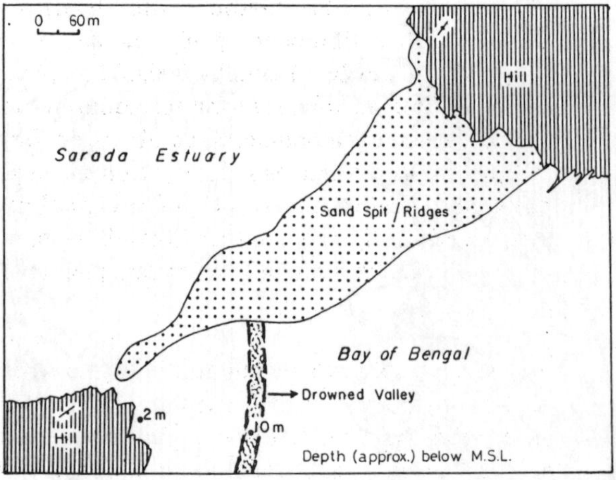

FIG. 27.5: Map showing the Drowned Valley near Sarada Estuary.

Discussion and Conclusions

The record of (glacial) low stands of sea level is submerged and largely inaccessible (Alt, 1968). The drowned valley near the Sarada estuary with a depth of about 10 metres together with the (natural) entrance channel to Visakhapatnam harbour having the same depth (of about 10 metres), the former cut across the structure of the rocks and the latter running along the structure indicate a conspicuous low stand of sea level between 10 and 11 metres (6 fathoms approx.). Though the alignment of the drowned valley near the Sarada estuary is along the NW-SE trend of the Dharwarian structural lineaments (Varadaranaj and Ganju, 1969) which control most of the master drainage channels in this region of the Eastern Ghats and the drowned channel of Narava Gedda (harbour entrance channel) along a probable fault (Natarajan and Nanda, 1981), the influence of changing base levels in the cutting of these valleys cannot be ruled. An uninterrupted, uniform and steep topographic (cliff) face up to a depth of 6 fathoms indicates an equilibrium profile of cyclic development with sea level standing stable for a considerable period of time. The sea level though fluctuating, has remained reasonably fixed during the last few thousand years; and it may well have been at least equally constant in the not very remote past (in an interglacial age or ages) for very much longer periods (Cotton, 1954). This near shore profile of the sea bottom is another piece of supportive evidence for the 10-11 metres low stand of sea level in this region.

Acknowledgements

The authors are grateful to Prof. R. Vaidyanathan, Deptt. of Geography, Andhra University, Visakhapatnam for his help during the course of the work and also for his valuable suggestions during the preparation of this paper. Thanks are due to the Head, Deptt. of Geography, Jamia Millia Islamia, New Delhi, for providing facilities. The authors are specially grateful to the DST, New Delhi, for granting a project, as a part of which this piece of work is undertaken.

REFERENCES

Alt, D. (1968): Pattern of Post Miocene eustatic fluctuation of sea level, *Jour, Palaeogeog. Palaeoclim Palaeoeco.*, Vol. 5, pp. 87-94.

Banerjee, M. and Sen, P.K. (1987): Palaeobiology in understanding the change of sea level and coastline in Bengal basin during Holocene period, *Indian Jour of Earth Sci.*, Vol. 14, pp. 307-320.

Chappel, J. and Shackleton, N.J. (1986): Oxygen isotopes and sea level, *Nature*, Vol. 324, pp. 137-140.

Cotton, C.A. (1954): Tests of a German non-cyclic theory and classification of coasts, *The Geog. Jour.*, Vol. 120, pp. 353-361.

Fairbridge, R.W., (1960): The changing level of the sea, *Scientific American*, Vol. 202, pp. 70-79.

——— (1961): Eustatic changes in sea level (in) *Physics and Chemistry of the Earth*, Pergamon Press, Vol. IV, pp. 99-185.

Kale, V.S. and Rajaguru, S.N. (1985: Neogene and Quaternary transgessional and regressional history of the west coast of India: An overview, *Bull. Deccan College Res. Instt.*, Vol. 44, pp. 153-167.

Mahadevan, C. and Poornachandra Rao, M. (1954): Study of ocean floor sediments off the east coast of India, *Andhra Univ. Memoirs in Oceanography 1*, pp. 1-35.

Moore, W.S. (1982): Late Pleistocene sea level history (in) *Uranium Series Disequilibrium: Applications to environmental problems*, Clarendon, Oxford, pp. 481-495.

Murthy, K.S.R., Rao, I.C.S and Sarma, K.V.L.N.S. (1987): Sub-bottom profiling over the inner shelf off Visakhapatnam, East Coast of *India, Indian Jour of Earth Sciences*, Vol. 14, pp. 109-113.

Nair, R.R. and Qasim, S.Z., (1979): Occurrence of a Bank with living corals off the SW coast of India, 'Mahasagar', *Indian Jour. of Marine Sci.*, Vol. 7, pp. 55-58.

Natarajan, V. and Nanda, J.K., (1981): Large scale basin and dome structures in the high grade metamorphics near Visakhapatnam, *South India, Jour. Geol. Soc. India*, Vol. 22, pp. 584-592.

Patel, M.P., Patel, S.G. and Merh, S.S. (1984): Geomorphic evidences of Quarternary sea level changes in Mahi-Tapi coastal segment of Gujarat, Proc. Symp. Quarternary Episodes, Deptt. of Geology, M.S. Univ., Baroda, pp. 49-62.

Prudhvi Raju, K.N. and Vaidyanathan, R. (1978): Geomorphology of Visakhapatnam, Andhra Pradesh, *Jour. Geol. Soc. India*, Vol. 19, pp. 26-34.

Rao, T.C.S., Machado, X.T. and Murthy, K.S.R. (1980): Topographic features over the continental shelf off Visakhapatnam, 'Mahasagar', *Indian Jour. of Marine Sciences*, Vol. 13, pp. 23-28.

Sadakata, N. (1980): Boring data and fossil pollen analysis in the Cauvery delta, South India, Chapter 9 (in) K. Fujiwara (ed.), Geographical field research in south India, Spl. Pub. No. 8, Res. and Sources Unit for Reg. Geog., Univ. of Hiroshima, pp. 175-179 (in) Meijerink, A.M.J., *Dynamic geomorphology of the Mahanadi delta, Itc Jour.*, 1982-83, pp. 243-250.

Shackleton, N.J. and Opdyke, N.D. (1973): Oxygen Isotope and Palaeo magnetic stratigraphy of Equatorial Pacific Core V 28, 238, *Jour. Quat. Res.*, Vol. 3, pp. 39-55.

Tanner, W.F. (1968): Introduction—Tertiary Sea Level Symposium, *Jour. Palaeogeog. Palaeoclim. Palaeoeco.*, Vol. 5, pp. 7-14.

Varadarajan, K. and Ganju, J.L. (1969): Lineament analysis of coastal belt of Peninsular India, *Memoirs Geol. Survey of India*, No. 12, pp. 49-58.

28

The Longitudinal Profiles of the Selected River Basins of Rajasthan

ABHA MATHUR

Introduction

The physiography of Rajasthan is of diversified nature and it is due to various disturbances during its geological history. Being a region of relatively moderate slope, the flowing water is drained by numerous rivers, some flowing across it whereas some originating and flowing through it. The rivers are nothing but the result of the nature of the terrain and the climate of the area. The movement of water depends upon the degree of the slope of the drained area. Hence the origin of the rivers and their intensity of flow are merely the result of the physiography of the area.

Objective

The objective of this paper is to study the longitudinal profiles of the selected rivers of Rajasthan, with a view to understand its physiography from one of the major criterians.

Definition

The longitudinal profile of a river implies 'The outline of the shape of a river valley from source to mouth'. To the Geomorphologists, it is well-known that the nature of a river profile gives a clue to the stage of the terrain in the framework of the cycle of erosion. Thus the specific type of river profile is both both an outcome of the stage of the terrain and an indication to the stage of the cycle of erosion.

The long profiles with knickpoints with represents some upliftment or rejuvenation in the life of the area. The long profile with high gradient and steep curve exhibit the young stage of the terrain whereas as the long profile with smooth curve (the logarithm curve) represent the long life and old stage of the terrain.

Thus the study of the longitudinal profiles help us in analysing the :

(i) Velocity of water flow,
(ii) Geological history of the terrain drained by the river.

Discussion

In the present study, the longitudinal profiles of 9 major selected rivers have been drawn. These river are as the following:

(1) The Kantli (Fig. 28.1)
(2) The Sabi (Fig. 28.2)
(3) The Chambal (Fig. 28.3)
(4) The Banganga (Fig. 28.2)
(5) The Banas (Fig. 28.2)
(6) The Mahi (Fig. 28.4)
(7) The Western Banas (Fig. 28.2)
(8) The Berach (Fig. 28.1).
(9) The Luni (Fig. 28.4)

1. The Kantli River

The river has three marked gradients (Fig 28.1):

(i) Between the source and Pachlangi, the gradient overages 23′ 6" per mile.
(ii) Between Pachlangi and ford, the gradient has decreased to a considerable extent and it values about 3′ 6" per mile and
(iii) Between ford and Kurrah, the gradient has again shown a marked increase over the middle profile of the river and this value average about 10′ 4′ per mile.

2. *The Sabi*

The Sabi, well-known for its badname as the sorrow of Rajasthan originates at an elevation of 2,000 feet in the Sawar Hills. Between Sawar Hills to the Sota wall, the gradient is 36: 3 feet per mile and thus the flow of the river is quite high and this high gradient is one explanation for the flood problem in this river basin. But between Sota nala and Siwari the gradient is only five feet per mile. One baffling question which arises here is why does the flood problems frequently arise despite a low gradient in the lower part of the river Sabi? The explanation is that the water flowing very speedily in its upper valley, loses its speed hence the river bed overflows. Secondly, the lowerbed is chocked with the sand deposits. As a consequence the flood problem has become a regular feature of this river.

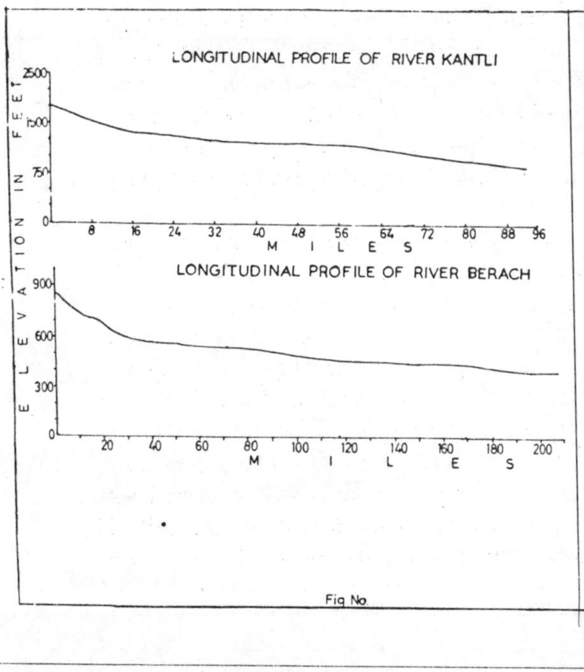

FIG. 28.1: Longitudinal Profile of River Kantli.

3. *The Chambal, Banas and the Mej*

As Fig. 28.3 very vividly displays that the profiles of these rivers are marked with knickpoints. The Chambal has three knick points at the elevations of:

(i) About 225 mt,
(ii) About 275 mt,
(iii) 360 metres.

Similarly the longitudinal profile of the Banas also represents three marked knickpoints at the elevations of about :

(i) 1600 feet,
(ii) 1250 feet, and
(iii) 750 feet.

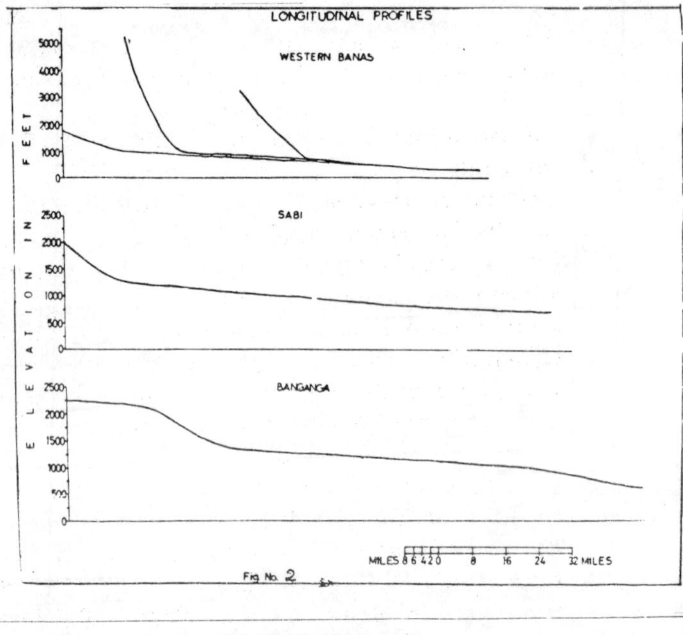

FIG. 28.2: Longitudinal Profiles : Western Banas.

The Parbati and the Kali Sindh represent their knickpoints at the elevation of about 230 metres and 270 metres respectively.

Thus these rivers strongly favour the opinion that this part of Rajasthan has undergone upliftment in its geological history and these rivers have consequently been affected. This is why these rivers are characterised with the geomorphic features of rejuvenation, i.e., the high gradient, the high velocity and the high rate of valley deepening.

The average gradient of these rivers are as given in Table 28.1.

TABLE 28.1: The Average Gradient of Rivers

Sl. No.	Name of the river	Average gradient in m. per km
1.	The Chambal	0.75
2.	The Banas	1.09
3.	The Mahi	1.9
4.	Kalisindh	1.36
5.	Kunu	2.08

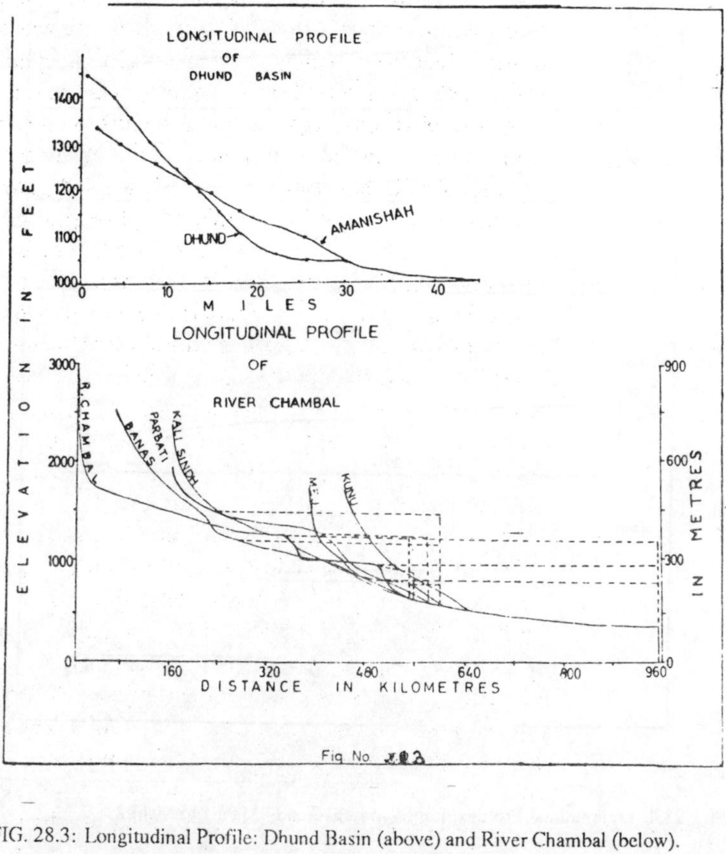

FIG. 28.3: Longitudinal Profile: Dhund Basin (above) and River Chambal (below).

4. *The Banganga*

The river at an elevation of about 1852 feet, has a knickpoint. It is here that the Duir Nallah joins this river. Afterwards, the profile of the river appears to be smooth which indicates the mature stage of the river and the geological stability in the eastern plain of Rajasthan. The knickpoint in the Aravallis represent the upliftment of the Aravallis during the cretaceous period. The average gradient of the profile is about 10 ft per mile.

—

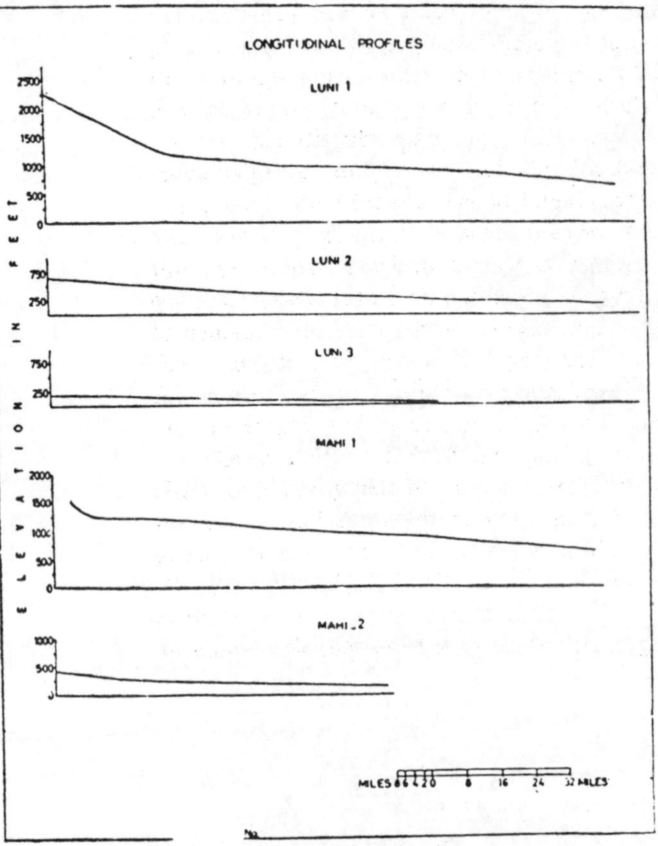

FIG. 28.4: Longitudinal Profiles: Luni 1 to Luni 3 and Mahi 1 to Mahi 2.

5. *The Mahi*

The Mahi's longitudinal profile resembles a smooth curve, with uniform gradient of the value of 6.5 feet per mile. The longitudinal profile favours that this area of Rajasthan has been stable when the Aravallis came into existence (Fig. 28.4).

6. *The Western Banas*

This river originally originates from an elevation of 1750 feet and when it enters into Gujarat, the elevation decreases to 250 feet. But it is at the elevation of about 1,000 feet, that the profile shows a marked change in its gradient. Similarly, at the same elevation, the Sukli which is a tributary to the western Banas, is also being characterised by a knickpoint. Thus these two knickpoints strongly plead the case of the upliftments of the adjoining area during the period when the Aravallis were formed (Fig. 28.2).

Sili, a sub-tributary to the western Banas, also represents a knickpoint at an elevation about 750 feet.

The gradient of the western Banas as a whole is about 15 feet per mile and the profile below the elevation of 1,000 feet is of smooth nature.

Conclusion

Rajasthan has 59 river-basins from which 9 important basins were chosen representing whole of the area to study the geological history in the framework of erosion. The longitudinal profiles of the rivers reveals that there was a sequence of upliftment in the geological history of the region. The profiles exhibit knickpoints and thus support the view of erosion surfaces.

29

Morphometric Characteristics and Net Sown Area in the Khari River Basin, Rajasthan

S.C. KALWAR and P.L. VIJAY

MORPHOMETRY MAY be defined as the measurement and mathematical analysis of the configuration of the earth's surface and of the shape and dimensions of the land forms. Agricultural activities depend on physical environment e.g. rugged relief determines the farm practices. Similarly it is universal fact that with the increase in steepness of slope the use even very simple farm machinery becomes difficult.

The Study Region

The Khari, a left bank tributary of the Banas river, originates from Deogarh and Kumbhalgarh hills. This basin is situated in the south-eastern part of Rajasthan State and spreads over in Deogarh and Bhim tahsils of Udaipur district, Asind, Hurda, Mandal, Banera, Shahpura tahsils of Bhilwara district and Beawar and Kekri tahsils of Ajmer district. This basin stretches between 25°27′ N to 26°7′30" N and 73°5" E to 75°7"30" E covering an area of 6,133 sq kilometres (Fig. 29.1). The length of the basin from north to south is 65.9 kilometres and width from east to west is 152 kilometres. The shape of the basin is elongated.

Table 29.1 and Fig. 29.2 show district, tahsils and Inspector Land Record (ILR) Circles of the Khari Basin.

The eastern part of the basin belongs to lowland where the height is below 375 metres. The central and western uplands are also almost flat and covers approximately one half area of the basin. The western part belongs to western upland, highland and high rugged terrain, the height varies from 450 m to 955 metres. The highest peak of the basin also lies in this part, i.e., Kumbhalgarh ghat peak (955 m).

There are two seasonal rivers,viz. the Khari and the Mansi river. The Khari river, flows from west to east. There are several *nallahs* join the river on the left bank, while the Mansi and other two *nallahs* join the river on the right bank.

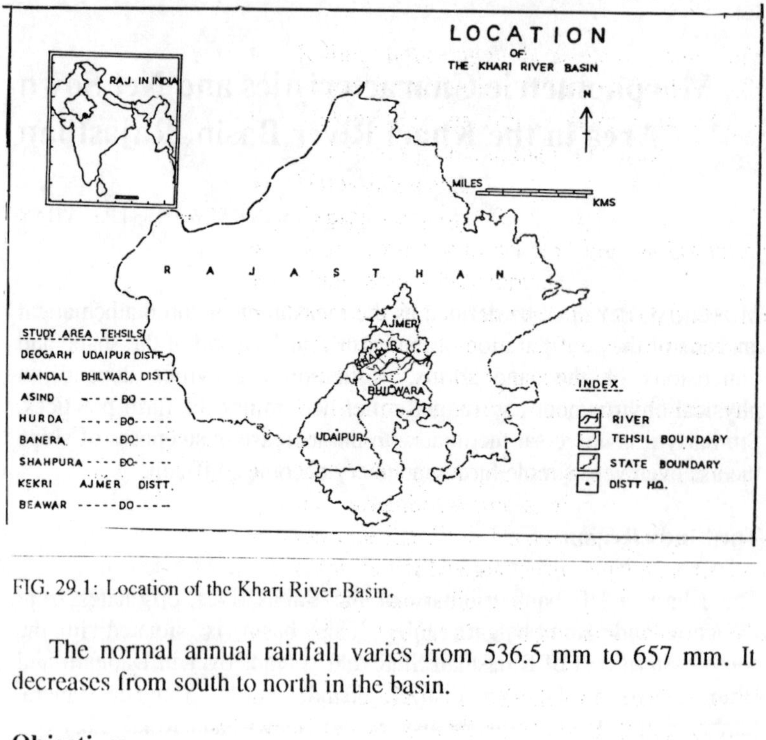

FIG. 29.1: Location of the Khari River Basin.

The normal annual rainfall varies from 536.5 mm to 657 mm. It decreases from south to north in the basin.

Objectives

The objectives of the study are :

(i) To analyse the morphometric properties of the Khari basin.
(ii) To study interrelationship of morphometric properties and net sown area.

Methodology

The present study is based on computation of morphometric data from

1: 50,000 topographical sheets of the Khari Basins. The completed data have been presented on log graphs against net sown area. Landuse data have been collected from revenue record office of Inspector circles.

TABLE 29.1: District, tehsils and ILR circles in the Khari Basin, 1988-89

Sl.No.	District	Tahsil	ILR Circle	Area in square kilometres
1.	Udaipur	Deogarh	Deogarh	24780
2.	Bhilwara	Asind	Asind	286.28
			Kareda	330.26
			Daulat garh	291.33
			Badnor	248.04
			Kalivas	307.95
		Hurda	Hurda	222.60
			Dantra	293.25
			Rupaheli	181.45
		Banera	Banera	234.51
			Upreda	209.60
			Decola	288.48
		Shahpura	Shahapura	279.13
			Arwad	284.36
			Phooliyakalan	279.40
			Deoliakalan	206.60
3.	Ajmer	Kekri	Nagola	235.26
			Kekri	285.60
			Khawas	261.19
			Sawar	217.18
			Sarwar	216.80
		Beawar	Masuda	231.90
			Ramgarh	161.72
			Bijai Nagar	243.43

Source: Land records, Udaipur, Bhilwara and Ajmer districts.

Morphometric Analysis

In morphometric analysis stream frequencies, drainage density, altimetric frequency histogram and slope analysis are included.

Stream frequencies

The stream frequency ranges between 0.04 to 0.80 . The maximum stream frequencies are found in western part of the basin where most of the area falls under 0.40 to 0.80 stream per unit area. The stream frequencies has been divided into three zones (Table 29.2).

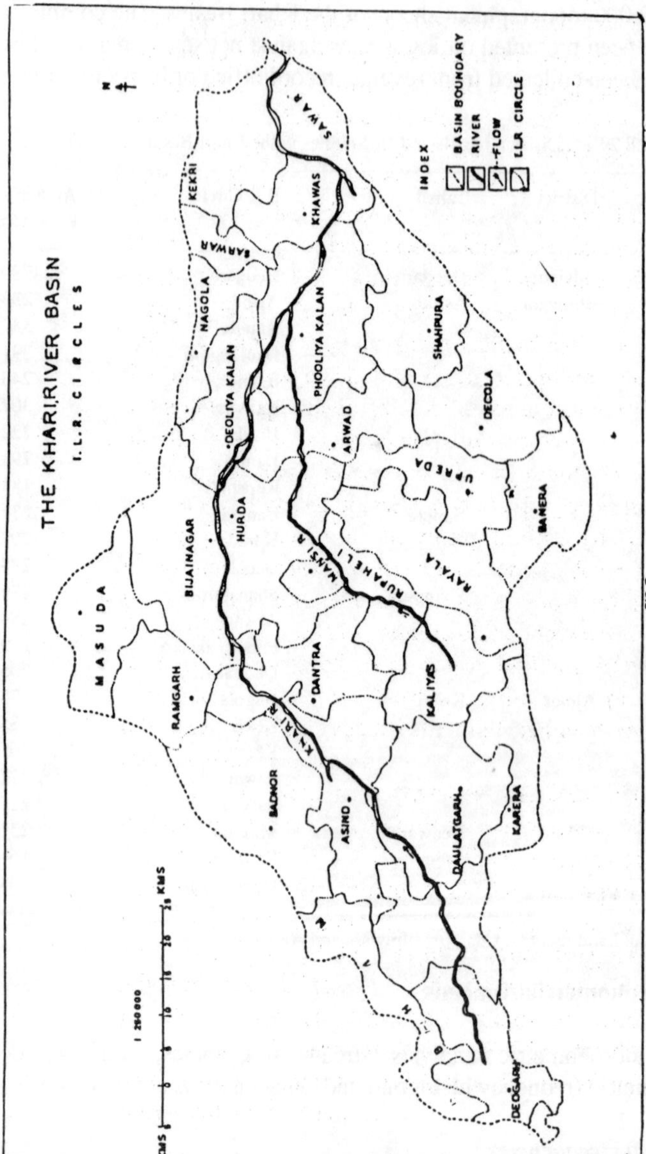

FIG. 29.2: The Khari River Basin: ILR Circles.

TABLE 29.2: Zones and Stream Frequencies

Zone	Frequencies	% of Area of basin
I The zone of low frequency	0.00 - 0.80	60
II The zone of medium frequency	0.08 - 0.40	37
III The zone of high frequency	0.40 - 0.80	3

Table 29.2 reveals that about two-thirds area is under low stream frequency zone while only 3 per cent area has high stream frequencies.

Drainage Density

The Khari river basin has divided into three drainage density zones as shown in Table 29.3.

TABLE 29.3: Drainage Density Zones

Zone	Drainage density per km^2	Area of basin in percentage
Maximum density zone	0.30 above	3
Medium density zone	0.21 - 0.30	25
Low density zone	0.10 - 0.20	70

Maximum drainage density is concentrated in the western part while low drainage density zone lies in the central part of the basin.

Altimetric Frequency Histogram

The altimetric frequency histogram deals with the frequency of occurrence of different elevations groups in a region. This diagram is quantitative and objective in nature and can be objected to statistical analysis (Clark, 1967). The altimetric frequency histogram has been prepared in order to show the frequency of spot heights at different elevations.

The altimetric frequency histogram of the Khari River basin gives an idea of the relative positions and percentage of different land surfaces of the basin (Fig. 29.3). It is quite obvious from Fig. 29.3 that the land surface having 300-450 meters elevation is dominant in the basin. It covers approximately 53 per cent of the total geographical area of the basin. The western part of the basin is highly elevated where elevation is more than 900 metres (1% area of the basin). About one-fourth area has 450-600 metres elevations while one-fifth area is below 900 metres, i.e., 600-750 and 750-900 metres. Thus the area can be categorised as undulated.

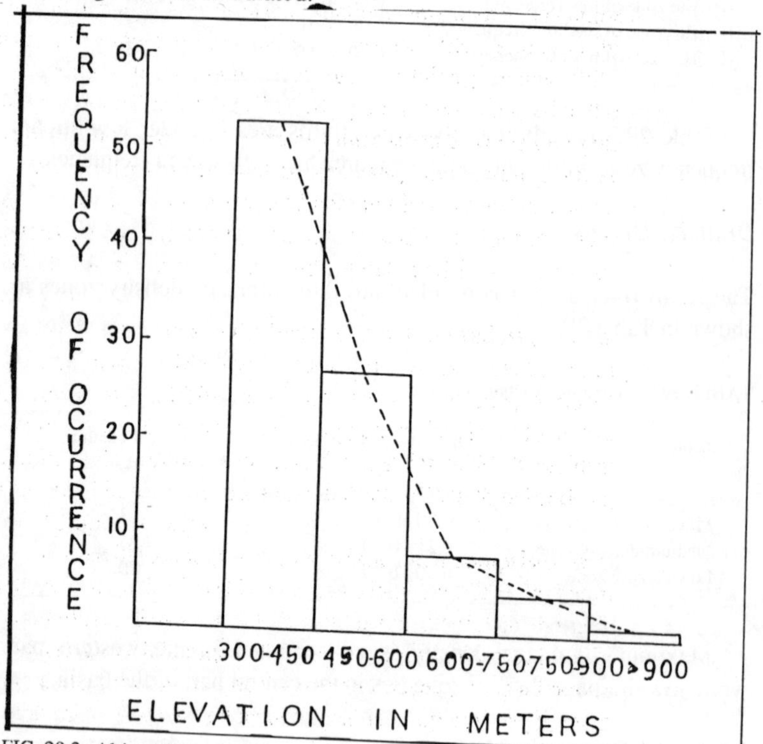

FIG. 29.3: Altimetric Frequency Histogram.

Slope Analysis

The slope analysis of the basin has been done by applying the Went Worth's Method (Tan = N/L . V1/3361).

The basin has an average slope of 0.08°. In northern and south-western parts of the basin, the slope is above 1°20′ Generally the slope decreases from west to east.

General Landuse Pattern

The area under forest is only 3.27 per cent to the geographical area.

The maximum concentration of forest is in the western part of the basin due to hilly terrain while the eastern part of the basin is peneplained where most of the area is under cultivation.

Land put to non-agricultural uses includes area under settlements, roads, railway tracks, reservoirs and parks etc. The Khari river basin covers 5.91 per cent to the total geographical area under this category. Barren and unculturable land covers barren and *'usar'* land which cannot be taken under cultivation except at a high cost. This type of land may be as isolated blocks within cultivated land. The Khari river basin has 16.17 per cent area under barren and culturable land.

Permanent pastures and other grazing lands are found mostly in Kareda (16.18%), Bhim (16.06%), Banera (16.06%), Deogarh (13.27%), Kaliavas (12.37%), Raila (13.30%) and Asind (12.08%) I.L.R. (Inspector Land Record) circles. These circles are situated in the western part of the basin, where most of the area is under hills or undulated. Only four I.L.R. circles viz. Masuda, Ramgarh, Bijainagar and Sarwar have less than ten per cent area under permanent pastures and grazing land (8.16 to 9.96%). These circles lie in the peneplained eastern part of the Khari river basin. Old fallow land includes, land which was taken up for cultivation but is temporarily out of cultivation for a period of not less than one year and not more than five years. Under this category, the maximum area is in Bhim I.L.R. circle while it is lowest in Kekri circle. The high percentage of old fallow may be due to poverty of the cultivators, inadequate supply of water or less rainfall, sandy and saline soils and less remunerative farming.

Net Sown area (NSA) includes the land which is actually under crops during the current year. The N.S.A. is 34.71 per cent to the geographical area in the basin. Table 29.4 shows intensity of N.S.A. in various I.L.R. circles.

TABLE 29.4: Net Sown Area in the Khari River Basin

Percentage	ILR circles	No. of circles	Intensity
Above 40	Sarwar, Nagola, Deoliya, Kalan, Phooliya, Kalan Shahapura, Kekri, Khawar, Sowar.	8	High
31 - 40	Bijainagar, Ramgarh, Masuda, Arwad, Rupaheli, Hurda,	9	Medium
21 - 30	Deogarh, Badnor, Kaliavas, Dantra, Ralia, Banera.	7	Low
Less than 20	Bhim, Kareda	2	Very low

Table 29.4 reveals that the high percentage of N.S.A is eight I.L.R. circles which are situated in the eastern part of the basin. As started earlier that the eastern part is almost peneplained.

The central part of the basin particularly Bijainagar, Masuda, Dantra, Hurda, Asind, Badnor, Arwad, Sarwar, Ramgarh, Rupaheli, Kaliavas, Ralia, Dacola, Deogarh, Banera and Daulatgarh circles have medium and low percentage of the net sown area. The very low percentage of N.S.A. is only in two circles viz. Bhim and Kareda. the western part of the basin restricts the cultivation due to more hilly tracts and high percentage of barren lands. The N.S.A. increase from west to east.

During the field survey, it was observed that Manshi tributary of the Khari river has the highest percentage of N.S.A. due to profile soil and irrigation facilities.

Morphometric Characteristics and Net Sown Area

In the present study, an attempt has been made to find out the relationship between morphometric characteristics and net cultivated land in the Khari river basin. The study of slope is a very important aspect. It is a matter of common experience to find the operational cost rising with the increase in steepness and *vice versa*. Farmers will not prefer steeper slopes which were alternatives are available. Steeper slopes are generally avoided and ploughed only if warranted by population pressure (Jasbir Singh, 1987).

Table 29.5 shows the slopes in degrees and intensity of net sown area in the Khasi rive basin. This table reveals that Deogarh, southern part of the Bhim and eastern part of the Masuda ILR circles have 1°01′ to above 1°20′ slopes, i.e., high to very high in the Khari river basin due to hilly terrain, where the net cultivated land is low to very low, i.e., 20 or less than 20 per cent of the total geographical area. Contrarily the circles of high intensity of net cultivated area situated in the eastern peneplained area where the slope is less than 0°08′ to 1°0′ (medium) slope and medium intensity of net cultivated land (21 to 40%) are situated in the western part of the basin except Arward and Decola circles, where the relief is undulated and hilly. This basin is an excellent example of negative relationship between slopes and net cultivated land. The results of stream frequencies and drainage stream densities are almost identicals. Table 29.6 shows a comparative statement of stream density and net sown area in the Khari river basin.

TABLE 29.5: Comparative Statement of Slopes and Net Sown Area in the Khari River Basin

Slope in degree	Intensity	ILR circles	Area of basin (%)	N.S.A in %	Intensity of N.S.A
Above 1°20′	V. High	Eastern part of Masuda and Deogarh	1	Less 20	Very low
1° 01′ to 1° 20′	High	Western part of the Deogarh and southern part of Bhim.	2	20	Low
0° 08′ q°	Medium	Deogarh, Bhim, Badnor, Asind, Ramgarh, Kalivas Arwad, Decola	17	21 - 40	Medium
Less 0° 08′	Very Low	Kekri, Sarwar, Khawas, Nagola, Masuda, Bijainagar, Shahpura, Phoolivas Kalan, Upreda, Banera, Raila, Hurda and Kera	80	Above 40	High

Source: Computed by authors.

TABLE 29.6: Comparative Statement of Stream Density and Net Sown Area in the Khari River Basin

Stream density per/km²	Intensity	ILR circles	Area of basin (%)	Net sown area in %	Intensity
Above 0.30	High	Deogarh, Bhim,	5	Less 20	Very low
0.21 - 0.30	Medium	Sarwar, Masuda, Banera, Deocoda, Kalivas	25	21 - 40	Medium
0.10 - 0.20	Low	Kekri, Kareda, Huda, Dantra, Rupaheli, Upreda Sarwar, Bijainagar Asind Ramgarh, Shahpura, Ralia.	70	Above 40	High

Source : Computed by authors.

Table 29.6 reveals that the stream density ranges between 0.10 to 0.30 per km². The maximum stream density (more than 0.30 per km²) is found in the western hilly tract of the basin particularly in Bhim and Deogarh circles while the medium drainage density (0.21 to 0.30 per km²) is scattered in Sarwar, Masuda, Decola and Kaliavas circles. The low stream density (0.10 to 0.20 per km²) is found approximately in three-fourths area particularly in the eastern part of the basin.

The comparative statement reveals that the high stream density (above 0.30 per km²) is concentrated in Deogarh and Bhim circles where the net cultivated land is very low (less than 20%) due to ruggedness. The reverse trend can be seen in the eastern part of the basin where stream density (0.10 to 0.20 per km^2) is low while the net cultivated land is very high (above 40%) due to fertile low land. It is clear from the fact that there is a negative relationship between stream density and net sown area.

Values for determining the line of best fit by the method of least squares for drainage density (Dd) — net sown area (NSA), slopes—net sown area, and stream frequency — net sown area have been computed. The results are shown in Figs. 29.4 to 29.6.

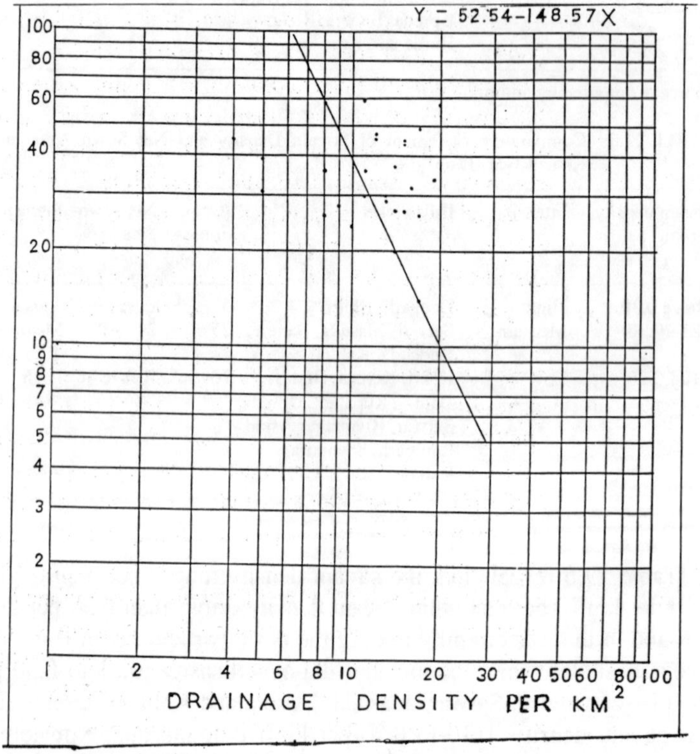

FIG. 29.4: Net Sown Area in Per cent: Dd/km^2.

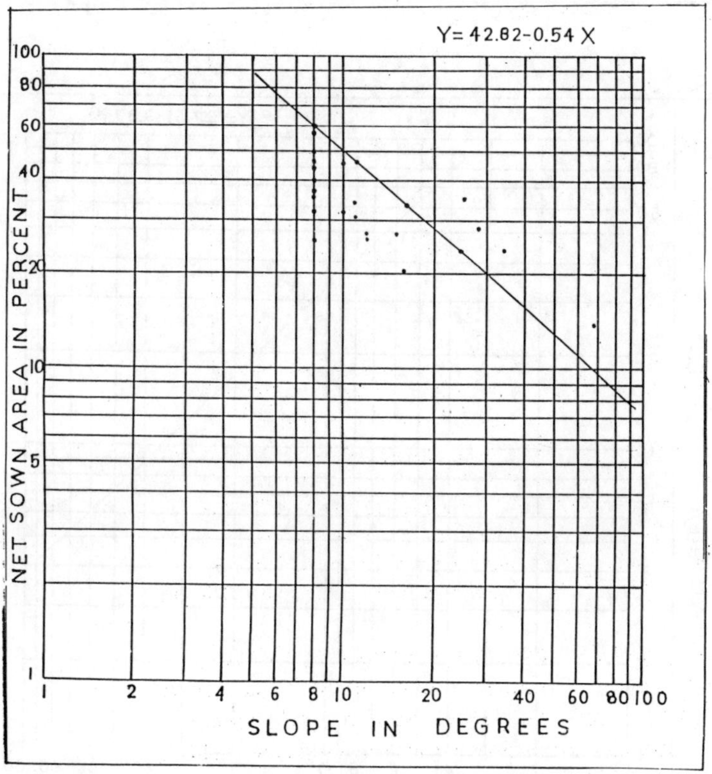

FIG. 29.5: Net Sown Area in Per cent: Slope in Degrees.

Conclusion

It may be concluded that the intensity of morphometric characteristics are decreasing from west to east while the net sown area (net cultivated) is increasing from west to east. Finally it can be summarised that there is a negative relationship between net sown area and morphometric characteristics e.g. stream frequency, stream density and slope.

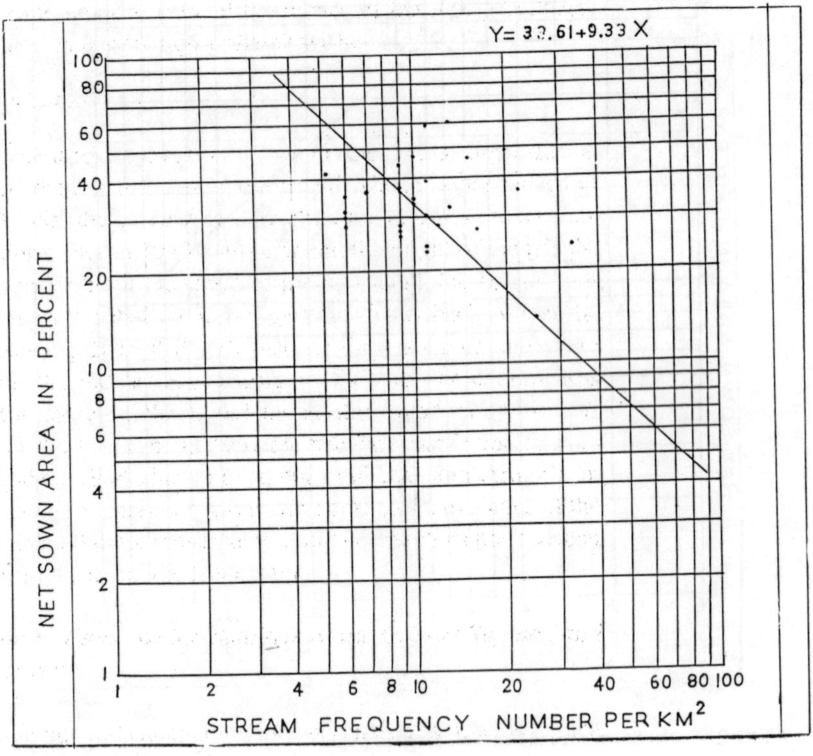

FIG. 29.6: Net Sown Area in Per cent Stream Frequency Number/km^2.

APPENDIX I

EXAMPLE OF COMPUTATION

Sl.No.	Name of the circle	$Y_{(NSA)}$	$X_{(Dd)}$	X^2	Y_x
1.	Deogarh	24.08	0.24	0.057	0.779
2.	Bhim	13.95	0.17	0.028	2.371
3.	Kakera	19.97	0.14	0.019	2.795
4.	Asind	34.14	0.19	0.036	6.486
5.	Badnor	27.67	0.09	0.008	2.490
6.	Kaliavas	27.44	0.14	0.019	3.841
7.	Hurda	30.43	0.13	0.016	3.955
8.	Dantra	29.66	0.13	0.016	3.855
9.	Rupaheli	32.26	0.16	0.025	5.161
10.	Raila	25.75	0.08	0.006	2.060
11.	Banera	26.72	0.15	0.002	4.008
12.	Upreda	26.35	0.17	0.028	4.479
13.	Decola	30.17	0.11	0.012	3.318
14.	Shahapura	46.65	0.12	0.014	5.598
15.	Arwad	36.05	0.08	0.006	2.884
16.	Phoolia Kalan	44.32	0.15	0.002	6.648
17.	Deoliya Kalan	45.94	0.12	0.014	5.512
18.	Nagola	45.51	0.04	0.001	1.820
19.	Kekri	59.66	0.11	0.012	6.652
20.	Khawas	59.00	0.08	0.006	4.720
21.	Sawar	49.42	0.17	0.028	8.401
22.	Daulat Garh	23.60	0.10	0.010	2.360
23.	Masuda	36.47	0.01	0.012	4.001
24.	Ramgarh	31.42	0.05	0.002	1.571
25.	Bijai Nagar	37.03	0.14	0.019	5.184
26.	Sarwar	39.04	0.05	0.002	1.952
	Total	902.70	3.22	0.388	106.240
	Mean	34.72	0.12	0.014	4.086

REFERENCES

Durry, G.H. (1970): *Essay in Geomorphology*, Heinmann, London, p. 235.

Kalwar, S.C. (1984): Changing Cropping Characteristics — A Case Study of Jaipur District. *Annals of the Association of Geographers*, Vol. IV, pp. 1-11.

Kind, C.A.M. (1970): *Techniques in Geomorphology*, Edward Arnold Pub. Ltd., London, pp. 1-17.

Leopold, L.B. (1969): *Fluvial process in Geomorphology*, Eurasia Publishing House Pvt. Ltd., Ram Nagar, New Delhi.

Leback, A.K., (1969): *Geomorphology*, McGraw-Hill Book Company, Inc., p. 3.

Morgan, W.G., and Munton, R.J.C. (1966) : 'Geographic Contributions to Land Use Planning', *Annals of the Association of American Geographers*, Vol. 26, p. 91.

Pandey, B.N., (1983): *Role of Science and Technology in Agricultural Land Utilisation*, S. Chand and Company, New Delhi.

Savigear, R.A.G. (1965) : 'A Technique for Geomorphological Mapping', *Annals of Association of American Geographers*, Vol. 55, pp. 514.

Singh, G.R. (1987) : Regional Imbalances of Food Crop Productivity in Rajasthan. Unpublished Ph.D. Thesis, University of Rajasthan, Jaipur.

Singh, Jasbir & S.S. Dhilon (1987) : *Agricultural Geography*, Tata McGraw-Hill Publishing Company Ltd., New Delhi.

Singh R.L. (1970) : 'Morphometric Analysis of Terrain', Presidential address—Indian Science Congress Association, Hyderabad, pp. 2-3.

Wooldridge S.W. & Morgan, R.S. (1956) : *An Introduction of Geomorphology*, Longmans, pp. 21-29.

INDEX